P. M. Dew
K. R. James

Introduction to Numerical Computation in Pascal

Springer-Verlag New York Inc.

P. M. Dew
K. R. James
Department of Computer Studies
University of Leeds

First published 1983 in the United Kingdom by
THE MACMILLAN PRESS LTD
Sole distributors in the USA and Canada:
Springer-Verlag New York Inc.
175 Fifth Avenue
New York, NY 10010
USA

Printed in Hong Kong

Library of Congress Cataloging in Publication Data
Dew, P. M.
　Introduction to numerical computation in Pascal.

　Bibliography: p.
　Includes index.
　1. Numerical analysis—Data processing. 2. PASCAL
(Computer program language) I. James, K. R.
II. Title.

QA297.D48　1983　　519.4′028′5424　　82-19651

ISBN 0-387-91216-9　Springer-Verlag New York Inc.

Contents

Contents

*Sections marked with an asterisk may be read superficially without loss of continuity on first reading.

Preface

Our intention in this book is to cover the core material in numerical analysis normally taught to students on degree courses in computer science. The main emphasis is placed on the use of analysis and programming techniques to produce well-designed, reliable mathematical software. The treatment should be of interest also to students of mathematics, science and engineering who wish to learn how to write good programs for mathematical computations.

The reader is assumed to have some acquaintance with Pascal programming. Aspects of Pascal particularly relevant to numerical computation are revised and developed in the first chapter. Although Pascal has some drawbacks for serious numerical work (for example, only one precision for real numbers), the language has major compensating advantages: it is a widely used teaching language that will be familiar to many students and it encourages the writing of clear, well-structured programs. By careful use of structure and documentation, we have produced codes that we believe to be readable; particular care has been taken to ensure that students should be able to understand the codes in conjunction with the descriptive material given in the book.

Our aim, unusual in an introductory text, is to explain in some depth good-quality mathematical software, paying full attention (subject to the overriding aim of clarity) to robustness, adaptability and error control as well as to the theory underlying the methods. The central theme is the description of a Pascal mathematical library, mathlib, which we have designed primarily for teaching purposes and used for computing courses at Leeds University. This gives us a well-defined framework for the development of theory and algorithms, exposes some important problems of writing mathematical software, and provides a consistent thread of motivation and reinforcement. Many of the exercises and projects make use of the library or require the student to modify or extend the library codes.

The book can also be used as a source of up-to-date mathematical routines in Pascal which are not readily available elsewhere. It is hoped that this will make it useful to practitioners and hobbyists with Pascal on their own micros. All the routines have been tested on the Amdahl 470 V/7 and DECsystem-10 computers at Leeds University, and library subsets have been mounted successfully on an Apple microprocessor. In the interests of portability, we have adhered in our coding as far as possible to the Pascal standard defined by Jensen and Wirth (portability of the library is discussed in appendix D). As a teaching aid, versions of the library are available for mainframe computers and for some microprocessors (further information is given below).

A particular debt is due to each of the following, who read all or part of the manuscript, in some cases more than once, and gave us many valuable comments: Dr W. Riha, Dr I. Gladwell, Dr D. Sayers, Dr J. K. Reid, Dr Mary Almond, Dr Ruth Thomas, Mr K. Hopper. The computer printouts used in the book were produced with the kind co-operation of Dr A. A. Hock and Miss C. Redford of the University of Leeds Computing Service. It is a pleasure also to record our thanks to the diligent typists in the Department of Computer Studies at Leeds University, Audrey Abbott and her helpers.

Leeds, 1982

P.M.D.
K.R.J.

NOTE ON AVAILABILITY OF THE MATHLIB LIBRARY

The mathlib library as described in this book is available for distribution on mainframe computers. A version of the library has also been written to run under UCSD Pascal on the Apple II microcomputer. It is planned to make the library available on other microcomputers in the future. Further details and distribution costs can be obtained by writing to the Editor of the Computer Science Series at Springer-Verlag, New York.

Introduction

The aim of this book is to develop the mathematical and computer programming skills required to understand and write reliable, efficient programs for some of the more important mathematical computations. We refer to a collection of programs of this type as *mathematical software.*

The need for mathematical software is evident if we consider that, to investigate phenomena in the physical or social sciences, it is necessary to represent systems by mathematical models. For example, dynamical systems and electrical networks are often modelled in terms of differential equations. An essential feature of any such model is that it is only an approximation to the real-life situation; the investigator tries to discover the most significant terms and include these in his model. Even so, it is rarely possible to obtain an exact solution. Instead, the problem is normally reduced to a number of basic mathematical tasks, for example, the solution of simultaneous equations or the evaluation of definite integrals, which can be solved *numerically* to a specified accuracy. It is desirable to provide the modeller with a set of standard programs to perform these basic computations reliably and efficiently.

From the earliest days of computing it has been recognised that there is a need for *library routines* to perform the basic mathematical computations. Originally, a library was a collection of programs written by users and thought to be of general interest. The programs were expected to run correctly only in the particular computing environment of the installation in which they had been developed. As there are many different computer systems in operation and the lifetime of any one system is comparatively short, this was found to be very wasteful of time and effort.

Today there are a number of projects that aim to provide comprehensive collections of mathematical software in a form which is, as far as possible, independent of the computing environment (machine-independent). The objective is to develop routines which can be used on a variety of computers, compilers and operating systems, and to supply users with well-defined program specifications so that they can employ the programs sensibly and correctly. For example, in Britain in the early 1970s the Numerical Algorithms Group (NAG) was established, initially to provide British universities with a set of well-tried mathematical routines on the ICL 1900 series of computers. Subsequently the NAG library has been made available on a number of machine ranges and is sold as a commercial product.

A library of routines, regularly updated and maintained, has the great advantage that the latest developments in theory and practice can be passed on without requiring from the user any specialised knowledge of the subject. Indeed, a modern computer library of mathematical routines represents the 'state of the art' in numerical computation.

Much of the mathematical software used in practice is written in FORTRAN 66. One reason is that programs written in this language are portable; that is, they can be used on many computer systems with little modification. FORTRAN, however, does not lend itself easily to structured programming and is probably not a good vehicle for teaching programming skills[†]. Since our aim is to present programs that are well-structured and (we hope) easy to read, we use the modern programming language Pascal.

We shall consider a number of mathematical problems that are of basic importance, develop the underlying theory and, in the light of this, discuss how to write suitable library routines to perform the necessary computations. The routines are collected together in a Pascal library called mathlib, a summary of which is given in appendix B. The intention is that the reader should be able to understand as well as use the routines and, for this reason, some of the finer detail which would be included in a commercial library has been simplified in the interests of clarity.

[†]FORTRAN in its essentials is one of the oldest programming languages. More recent versions, for example FORTRAN 77, contain some facilities for structured programming.

Part 1

FUNDAMENTAL TECHNIQUES

Programming in Pascal

Part 1

FUNDAMENTAL TECHNIQUES

1 Programming in Pascal

Every number is a correct program, though perhaps for the wrong task.

Norman H. Cohen, *Gödel Numbers: A New Approach to Structured Programming*

In this chapter the main features of Pascal which will be used throughout the book are summarised. It is assumed that the reader has some acquaintance with the elements of Pascal programming (there are useful introductory texts, for example, by Findlay and Watt (1978) and Wilson and Addyman (1978)). Our programs will as far as possible follow the standard defined in the Pascal User Manual and Report by Jensen and Wirth (1978).

A basic convention of Pascal is that all programs have the same underlying structure. The structure of a program is

```
PROGRAM <name> ( list of program parameters ) ;

CONST
    <definition of any constants used in the program> ;

TYPE
    <definition of any nonstandard data types> ;

VAR
    <declaration of variables with type specifications> ;

<declaration of any subprograms, i.e. functions and procedures> ;

BEGIN
<body of program>
END .
```

1.1 BASIC DATA TYPES

Variables and constants used in a Pascal program must be of specified type. The type of variable is specified in a VAR declaration at the head of the program (or subprogram — see section 1.3). Pascal provides a number of standard data types, for example, integer, boolean and real, and also permits the programmer to define his own data types. Indeed, one of the attractive features of Pascal compared with other languages is the richness of data types which the user is allowed to define.

3

Integer

Type integer enables the programmer to represent and manipulate whole numbers; integer values are stored exactly in the computer and the elementary arithmetic operations +, −, * are performed exactly on such numbers (for division see under real numbers below.) The set of integer numbers which can be represented on a computer is restricted to the range

$$-\text{maxint}, \ldots, -1, 0, 1, \ldots, \text{maxint},$$

where maxint is a predefined Pascal constant. The value of maxint depends on the word length of the computer; for example, on the Amdahl V/7 it is 2 147 483 647. Any attempt to compute or to store an integer value lying outside this range normally results in an error message *integer overflow* and termination of execution.

The main use of integer variables in mathematical computing is for counting, controlling iteration loops and indexing array elements. Particularly useful in this context is the FOR statement

```
FOR i := m TO n DO
   BEGIN
   <Pascal statements>
   END
```

where i, m and n of type integer†. The BEGIN and END are optional if there is only one statement to be executed. The statements in the FOR loop are executed with $i = m, m + 1, \ldots, n$ (if $m > n$ the statements are not executed at all); m and n may be replaced by expressions of the appropriate type.

We often require to count backwards (that is, with $i = n, n - 1, \ldots, m$), and for this purpose we have the complementary statement

```
FOR i := n DOWNTO m DO
```

Boolean

A variable or expression of type boolean can assume the values true or false. We frequently encounter boolean expressions in conditional and looping structures. The reader should be familiar with the IF...THEN...ELSE construction and also with boolean expressions of the form

```
x < 0 ,      (x < 0) OR (y >= 0)
```

The use of parentheses in the latter is compulsory in Pascal, unlike some other programming languages.

In addition to the FOR statement, Pascal provides two other useful iteration constructs. One of these is the WHILE statement:

†Or in general any type which can be mapped into the set of integers (so-called ordinal types).

```
WHILE <boolean expression> DO
   BEGIN
   <Pascal statements>
   END
```

The BEGIN and END are optional if there is only one statement to be executed. The statements in the WHILE loop are executed repeatedly as long as <boolean expression> is true. The condition is tested before the start of each iteration and it is therefore possible for the statements not to be executed at all. In numerical programs we sometimes require a first iteration to be performed before a test is made, and Pascal provides an iteration construct which does this:

```
REPEAT
   <Pascal statements>
UNTIL <boolean expression>
```

The statements bracketed by REPEAT. . .UNTIL are executed at least once and repeatedly until <boolean expression> is found to be true.

Appropriate use of the control structures FOR, WHILE and REPEAT. . . UNTIL can enhance the clarity of a program, as is shown in examples 1.1 and 1.2 below.

Real

In scientific computing we make extensive use of real (that is, non-integer) numbers to represent physical quantities such as spatial co-ordinates and time. For this purpose we have in Pascal the data type real. A real constant may be written either in the usual decimal form or in floating-point (scientific) notation. The latter is particularly useful for very large or very small numbers. For example, a program for quantum theory calculations requiring Planck's constant may have the constant definition

```
CONST
   h = 0.66256E-26 ;    (* Planck's constant *)
```

The symbol E−26 denotes 'times 10 to the power −26' and its use is clearly preferable to writing twenty-six leading zeros. We note that in standard Pascal it is compulsory to have at least one digit before and after the decimal point. This applies even to zero; the real zero is written as 0.0.

The set of real numbers, like the set of integer numbers, is restricted in range, although the range is much greater for reals than for integers (the maximum real number is 10^{20} to 10^{320} on mainframe computers). The elementary arithmetic operations on real numbers are $+, -, *, /$. The operator / can be applied to real or integer operands and yields a real result; thus $-5/3 = -1.6666. . . .$ The operator DIV applies only to integer operands and yields an integer result with any fractional part discarded; thus -5 DIV $3 = -1$.

There are important differences between real numbers in computing and real numbers in mathematics. On a computer there is a maximum and a minimum positive real number. Furthermore, because the word length of the computer is finite, only a finite number of significant digits can be represented (the equivalent of 7 to 14 decimal digits is usual). It follows that constants such as π, $1/3$ and even $1/10$ (on binary machines) cannot be represented exactly; more subtly, an error can be introduced whenever the result of a real arithmetic operation is stored. This has fundamental implications for numerical computation which will be discussed more fully in chapter 4.

In libraries it is usual to specify any required mathematical constants, such as π, to the highest precision of computers in current use ($\pi = 3.14159265358979$ to 15 significant decimal digits).

To illustrate the use of real numbers and looping structures in Pascal, we consider a standard problem of mathematics and describe one way of solving it (more efficient methods for this problem will be developed in chapter 8).

Example 1.1

Write a program to estimate the value of the integral

$$\int_0^1 \sin(x^2)\,dx$$

The following program estimates the integral using a simple numerical method to approximate the area under the curve. The range from 0.0 to 1.0 is divided into n equal subintervals (steps) of width h. A rectangle is fitted in each sub-interval with a height equal to the value of $\sin(x^2)$ at the mid-point of the step. The total area of the n rectangles is then taken as an approximation to the area under the curve.

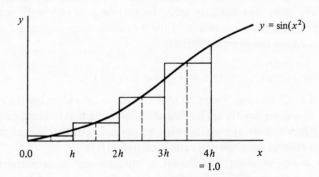

Figure 1.1   The mid-point (rectangle) rule with $n = 4$

```
PROGRAM integral ( input, output ) ;
(* This program computes an approximation to the integral of
   sin(x^2) from 0.0 to 1.0 using the mid-point rule with
   n steps *)

CONST
   uplim = 1.0 ;   (* upper limit of integration *)
   lolim = 0.0 ;   (* lower limit of integration *)
```

```
VAR
    (* Specify the types of all variables in the program *)
    i, n        : integer ;
    area, h, x : real ;

BEGIN   (* start of program body *)
read (n) ;                      (* number of steps required *)
h := (uplim - lolim)/n ;   (* width of each step *)
(* Accumulate the total area of n rectangles:  *)
area := 0.0 ;
FOR i := 1 TO n DO
    BEGIN
    x := lolim + (i-0.5)*h ;        (* mid-point of ith step *)
    area := area + h*sin(sqr(x))   (* area of ith rectangle *)
    END ;
writeln ('the mid-point rule with  n =', n) ;
writeln ('estimate of the integral is  ', area)
END .
```

The result is not exact for any finite value of n, but we would expect to be able to reduce the error by increasing n. Sample output from three runs of the program:

```
the mid-point rule with  n  =            4
estimate of the integral is   3.073851182027E-01

the mid-point rule with  n  =           40
estimate of the integral is   3.102401541321E-01

the mid-point rule with  n  =          400
estimate of the integral is   3.102680203152E-01
```

Notes

(i)   The CONST definition allows the values of uplim and lolim to be changed if different limits of integration are required. It is important to write our programs so that such constants can be easily located and modified.

(ii)   The statements in the FOR loop are indented to make the structure of the program clear, and comments are provided to assist in understanding the code. It is good programming practice to take care over layout and comments while the program is being developed.

(iii)   The *predeclared functions* sin and sqr are used to evaluate $\sin(x^2)$. A list of predeclared mathematical functions available in Pascal is given in appendix A.

(iv)   We see from the expression $(i - 0.5)*h$ that we are allowed to use mixed arithmetic, that is, expressions containing integer and real quantities. The expression is evaluated as real. We are also allowed to write assignment statements of the form $x := 1$, and conversion from integer to real is performed automatically. (The opposite is not the case. For conversion from real to integer we must use the functions trunc or round: see appendix A.) □

1.2 INPUT/OUTPUT

Pascal provides the standard files input and output which on most systems can be assigned at runtime to the user's terminal, named disk files or input/output devices as appropriate. The method of performing these assignments depends on the particular implementation and readers should consult the documentation for their computer systems. Additional files can be used, if required, by adding extra filenames to the program heading and inserting a filename as the first parameter in the read or write statements. The reader should consult a textbook on Pascal for further details.

Disk files are the most convenient means of handling large quantities of data. The use of an input data file avoids the tedium of typing the same data repeatedly while a program is being developed, thereby saving time and reducing the chance of error. Pascal provides a boolean function eof which is set to true when the end-of-file is reached. The use of this function, as shown in example 1.2, avoids the need to specify the length of a data file.

Example 1.2

Write a program to compute the root mean square of a set of real numbers. Assume that the input file is to be attached at runtime to a data file which contains the numbers.

```
PROGRAM rms ( input, output ) ;
(* This program computes the root mean square of a set
   of real numbers read from a data file *)

VAR
   n       : integer ;
   sum, x  : real ;

BEGIN
n := 0 ;       (* number of items read *)
sum := 0.0 ;   (* sum of squares *)
WHILE NOT eof DO
   BEGIN  (* read the data to end of file *)
   readln (x) ;
   n := n + 1 ;
   sum := sum + sqr(x)
   END ;
(* Print the number of items and root mean square *)
writeln (n:6, ' numbers in data file') ;
writeln ('    root mean square = ', sqrt(sum/n):12:6)
END .
```

For a file containing the data

        0.0
        0.5
        2.25
        5.0
        10.2

the output is

```
5 numbers in data file
root mean square =       5.183676
```

Notes

(i)   We input data with readln rather than read when using eof with a data file. The reason for this is rather technical (see Findlay and Watt (1978), section 17.3); the effect is that the reading position in the file is moved to the start of the next line when the value(s) in the readln list have been read. In our example this means that we should have one number on each line of the data file.

(ii)   The value of n in the writeln statement is output in a field of width 6 according to the format specification n:6. The output format of real numbers is controlled by a postfix of the form :w:d, for example, sqrt (sum/n):12:6. The effect is that the number is printed in a field of width w with d digits after the decimal point. If d is absent, the number is printed in floating-point form (scientific notation) in a field of width w. If both w and d are absent, default values are taken which depend on the particular compiler. □

1.3 SUBPROGRAMS

A subprogram may be regarded as a self-contained program module which, when supplied with suitable inputs from the main program or from another subprogram, performs a specified computation. The advantages of using subprograms are:

(i)   A problem is easier to conceptualise and hence to solve if it is first reduced to a number of subproblems, possibly through several levels of refinement. The eventual subproblems are often conveniently coded as subprograms.
(ii)   The high-level structure of the program and the interrelationship of its parts is made clearer; the program is therefore easier to understand and is less likely to contain programming errors.
(iii)   A subprogram can be 'invoked' (that is, caused to perform its computation) as many times as required from different parts of the program and with different input values; this avoids the need to repeat the same or similar pieces of code.
(iv)   A subprogram can be developed and tested separately from the main program and, once working correctly, can often be used in several different programs (this is the *raison d'être* of program libraries).

In Pascal a subprogram is either a FUNCTION or a PROCEDURE.

Functions

The predeclared functions sin, sqr, sqrt and eof have already been met in examples 1.1 and 1.2. Other functions are cos, abs, exp, ln etc. (see appendix A).

We now consider how to declare our own functions for use in a Pascal program.

Given a mathematical function such as

$$f(x) = ax^2 + bx + c$$

where a, b and c are constants, we may wish to calculate $f(x)$ for various values of the argument x. Once we have defined the function f, we can use it to evaluate $f(x)$ for specific values of x and also to compute more complicated expressions such as

$$\cos [f(x)] + \sin [f(1-x)]$$

A Pascal function is similar in effect to a function in mathematics. We must first *declare* (that is, define) the function, and we may then use it in our program with specified values of the argument(s), just as we use the predeclared functions sqr, sqrt etc. The declaration of a function takes the general form

```
FUNCTION <name> ( list of parameters, if any,
                         with type specifications )
               : <type of the function> ;

   <declaration of any local constants and variables> ;

   BEGIN
   <body of function>
   END
```

Example 1.3

Declare a Pascal function for

$$f(x) = ax^2 + bx + c$$

where a, b and c are constants. Incorporate the function in a program which defines the constants, reads in a value for x and computes

$$\cos [f(x)] + \sin [f(1-x)]$$

```
PROGRAM ftest ( input, output ) ;

CONST
   a = 1.0 ;   b = 2.0 ;   c = 3.0 ;
   (* the constants may be altered by editing the program *)

VAR
   xx : real ;

(* Declare the function *)
FUNCTION f ( x : real ) :   real ;
   BEGIN
   f := a*sqr(x) + b*x + c
   END (* f *) ;

BEGIN  (* main program *)
read (xx) ;
(* Print the value of the expression *)
writeln ( cos(f(xx)) + sin(f(1.0-xx)) )
END .
```

Notes

(i)   The FUNCTION declaration follows the CONST definition and the VAR declaration in the main program (this order is compulsory).

(ii)   The name of the function is f; it has one parameter, x, of type real, and the type of the function is also real.

(iii)   The body of the function contains a statement which assigns a real value to f; this is how the computed value is returned to the main program.

(iv)   The function is invoked twice from the main program with the *actual parameters* xx and 1.0 − xx, both of type real. A careful distinction must be made between an actual parameter and the *formal parameter* x in the FUNCTION declaration; x is the name of a variable which is given a value only when the function is invoked. (A formal parameter may sometimes have the same name as a variable in the main program but the two are entirely distinct.)

(v)   The constants a, b and c are *global*; that is, they are defined in the main program and their scope extends into the body of the function where they are used in the definition of f. □

Any constants or variables defined in the main program may be used within a subprogram, provided that their identifiers do not appear as formal parameters in the parameter list, and provided that they are not re-declared locally within the subprogram. The decision whether to use quantities global to a subprogram or to make them explicit as parameters is to some extent a matter of programming style. What we should do is to think very carefully before altering the values of global variables within a subprogram: this can make the working of the program difficult to follow and lead to unintended side-effects.

Procedures

The second form of subprogram in Pascal is PROCEDURE. This is distinguished from FUNCTION in that no type is given to a procedure and no value is assigned to the procedure name. Instead, certain values (usually more than one) are computed and returned when the procedure is invoked (the way in which this is done is described below). In fact, it is possible for a procedure to compute no values at all; we have had examples of this in the predeclared procedures read and write. These are special cases in that the number and types of the parameters can be varied; this is not the case with procedures which we declare ourselves.

A procedure is declared in a way similar to a function. The general form of the heading is

```
PROCEDURE <name> ( list of parameters, if any,
                   with type specifications )
```

The following example illustrates how to declare and use a procedure.

Example 1.4

Write a procedure to compute the roots (assumed to be real) of the quadratic
equation $ax^2 + bx + c = 0$, using the standard formula

$$[-b \pm \sqrt{(b^2 - 4ac)}] / 2a$$

Incorporate the procedure in a program which reads in the coefficients *a, b, c*
and prints the values of the two roots.

```
PROGRAM quadratic ( input, output ) ;

VAR
   a, b, c, x1, x2 : real ;

(* Declare the procedure *)
PROCEDURE quad (     a, b, c : real ;
                 VAR x1, x2  : real  ) ;

   VAR
      t : real ;  (* t is a local variable available only
                     within the body of PROCEDURE quad *)

   BEGIN  (* body of PROCEDURE quad *)
   x1 := 0.0 ;  x2 := 0.0 ;  (* default values of zero
                                if an error is detected *)
   t := sqr(b) - 4*a*c ;
   (* Check for an error in the coefficients *)
   IF t < 0.0 THEN
       writeln ('*** error in quad :  roots are complex')
       (* input/output statements are not normally used for
          error conditions in subprograms - see chapter 2 *)
   ELSE
      BEGIN
      t := sqrt(t) ;
      x1 := (-b + t) / (2*a) ;
      x2 := (-b - t) / (2*a)
      END
   END (* quad *) ;


BEGIN  (* main program *)
(* Read the coefficients and print them back as a check *)
read (a, b, c) ;
writeln ('the coefficients of the quadratic are: ') ;
writeln (a:16:6, b:16:6, c:16:6) ;
quad (a, b, c, x1, x2) ;  (* call quad to compute the
                             roots x1, x2 *)
writeln ('the roots are ', x1:16:6, ' and ', x2:16:6)
END .
```

Notes

(i)   The parameters of a procedure fall into two classes: *value parameters* (for
example, a, b, c) that are similar to the parameters of a function, and *variable
parameters* (for example, x1, x2) that are preceded by VAR in the parameter
list and have a special significance. When a subprogram is invoked, the value
parameters are initialised to the values of the corresponding actual parameters
in the calling program. Any subsequent change made to the value parameters
within the subprogram has no effect outside: value parameters are used only as

inputs to the subprogram. Variable parameters, on the other hand, stand for the corresponding actual parameters in the calling program (although there is no necessity for these to have the same name). Changes made to variable parameters within a subprogram change the values of the corresponding actual parameters. This is how results are passed back from the procedure. As a general rule, if a parameter is intended only as an input parameter and is of scalar type, for example, integer, boolean or real, then it should be a value parameter; otherwise, it should be a variable parameter preceded by VAR.

(ii)   To call/invoke a procedure, either in the main program or in another subprogram, we simply write the procedure name followed by a list of actual parameters, for example, quad (a, b, c, x1, x2). Actual parameters such as x1, x2 which correspond to variable parameters must, of course, be variables since the effect of the procedure may be to alter their values. Actual parameters that correspond to value parameters may be constants, variables or expressions of the appropriate type.

(iii)   Finally, we note that procedure quad suffers from a number of defects which make it unsuitable as a general-purpose routine. For example, it cannot compute complex roots and it does not test the value of the parameter a before attempting to perform a division; this would lead to overflow failure if a happened to be zero or 'small' (see also example 4.8). □

1.4 LIBRARY ROUTINES

So far we have seen that in Pascal, as in other programming languages, subprograms may enter computations in two ways:

  (i) as user-defined subprograms, for example, procedure quad in
      example 1.4,
  (ii) as predeclared functions and procedures, for example, sin, sqrt, write.

As an example of (i), a programmer who finds that the signs of real variables or expressions have to be determined several times in a program might declare

```
FUNCTION sign ( x : real ) :   integer ;
(* Returns the sign of x *)

   BEGIN
   IF x > 0.0 THEN
      sign := 1
   ELSE
      IF x < 0.0 THEN
         sign := -1
      ELSE
         sign := 0
   END (* sign *)
```

The sign function is of general utility in mathematical computations, and we would like to make this and other useful routines available to programmers in a way similar to the predeclared functions. We can do this by creating a Pascal library. A library is a collection of reliable and efficient subprograms which are designed to be useful in a particular applications area. In this book we shall develop a library of subprograms, called the mathlib library, for performing mathematical computations. We now see that in addition to (i) and (ii) above we may list a third source of subprograms:

(iii) as library routines, for example, routines from the mathlib library.

The mechanism for creating a library will be discussed in chapter 2. The specifications of the routines in mathlib, together with the definitions of library constants and data types, are given in appendix B. The first three routines of mathlib Group 1 (general utility routines) are summarised as

```
FUNCTION sign ( x : real ) :  integer ;
Returns the sign of x  (sign = 1 if x is positive,
= -1 if x is negative, = 0 if x is zero).

FUNCTION powerr ( x, r : real ) :  real ;
Computes x to the power r, where r is real.
Returns 0.0 in cases where x^r would underflow.
Writes an error message and causes runtime failure
if x^r is undefined.

FUNCTION poweri ( x : real ; i : integer ) :  real ;
Computes x to the power i, where i is an integer.
Returns 0.0 in cases where x^i would underflow.
Writes an error message and causes runtime failure
if x^i is undefined.
```

Pascal does not provide an operator or a predeclared function to compute powers. For inclusion in the mathlib library we write our own function (in fact two functions, since it is necessary to distinguish between exponents of type real and type integer).

To compute x^r, where r is real, we use the identity

$$x^r = \exp (r \ln x), \quad x > 0$$

($\ln$ denotes logarithm to base e; predeclared functions are available for exp and ln). We must be careful, however, to specify precisely the mathematical function we wish to program. When $x < 0$, the functions $\ln x$ and x^r are undefined in the real number system; on the other hand, when $x = 0$ and $r > 0$, we have $x^r = 0$ even though $\ln x$ is undefined. These cases can arise in the normal use of a library function and we must decide how to handle them. We classify the cases as follows

$$x^r = \begin{cases} \dagger\exp (r \ln x), & x > 0 \\ 0, & x = 0 \text{ and } r > 0 \\ ***, & x = 0 \text{ and } r \leqslant 0 \\ ***, & x < 0 \end{cases}$$

†A further refinement is implemented in the library functions powerr and poweri at this point. If the value is too small (that is, too close to 0.0) to be represented in the computer, Pascal will normally fail on *underflow*. We wish to prevent this from happening and return the value 0.0; see appendix C.

The symbol *** here indicates that x^r is undefined as a mathematical real number (x^r *is* defined for $x < 0$ and r an integer; however, as r is of type real it will not be possible to determine with complete certainty whether r is a whole number).

When x^r is undefined the library function powerr outputs an error message and forces runtime failure by attempted zero division. An alternative would be to return a default value and allow the computation to continue with this incorrect value. The action we have chosen is not ideal; in chapter 2 we shall develop a better way of handling error conditions in library routines.

When the exponent in the power function is of type integer we can compute the power directly by multiplication or division (function poweri). The possible cases are then

$$x^i = \begin{cases} xx\ldots x, & x \neq 0,\ i > 0 \\ 1, & x \neq 0,\ i = 0 \\ (1/x)(1/x)\ldots(1/x), & x \neq 0,\ i < 0 \\ 0, & x = 0,\ i > 0 \\ \text{***}, & x = 0,\ i \leqslant 0 \end{cases}$$

(A more efficient method than direct multiplication is used in the mathlib implementation — see appendix C.) There is one case where x^i is undefined; poweri handles this in a similar way to powerr.

Use of Library Routines

For completeness we give the code of the power functions powerr and poweri in appendix C. It is not, of course, necessary to see the code in order to use the library routines. Users of a library have access to 'black box' routines which they do not normally see. What users do see are the specifications which tell them how to employ the library routines, and it is therefore essential that these should be as clear and complete as possible (the documentation of library routines will be discussed in chapter 2).

The following example shows how two functions, sign and poweri, can be used from the mathlib library.

Example 1.5

Write a procedure to express a real number x in *normalised decimal floating-point form*

$$x = f \times 10^m, \text{ where } 0.1 \leqslant |f| < 1.0$$

(thus -0.0021 should be expressed as -0.2100×10^{-2}). Test the procedure on a range of values of x.

```
PROGRAM floatnumber ( input, output ) ;
(* This program reads a set of real numbers from a data file
   and writes them in normalised decimal floating-point form *)

VAR
   x, f : real ;
   m    : integer ;


(* Declare the library routines used in the program *)
FUNCTION sign ( x : real ) : integer ;
   EXTERN ;


FUNCTION poweri ( x : real ;
                  i : integer ) : real ;
   EXTERN ;


(* Declare a procedure to derive the normalised
   floating-point form, f*10^m, for a real number x *)
PROCEDURE float (      x : real ;
                  VAR f : real ;
                  VAR m : integer ) ;

   VAR
      signx : integer ;   (* local variable in float *)

   BEGIN
   m := 0 ;  (* initialise the exponent *)
   IF x = 0.0 THEN
      f := 0.0  (* the number is zero *)
   ELSE
      BEGIN
      (* Remove and save the sign of x *)
      signx := sign(x) ;  f := abs(x) ;
      (* Scale f to lie between 0.1 and 1.0 *)
      IF f < 0.1 THEN
         REPEAT (* scale f up *)
            f := 10*f ;  m := m - 1
         UNTIL f >= 0.1
      ELSE
         WHILE f >= 1.0 DO
            BEGIN  (* scale f down *)
            f := f/10 ;  m := m + 1
            END ;
      f := signx * f  (* restore the sign of x *)
      END
   END (* float *) ;


BEGIN  (* main program to test float *)
WHILE NOT eof DO
   BEGIN
   readln (x) ;
   float (x, f, m) ;  (* derive f*10^m from x *)
   writeln ;
   writeln (x:10:4, ' = ', f:7:4, ' * 10^', m:1) ;
   (* As a check, reconstitute x using poweri *)
   x := f * poweri(10.0, m) ;
   writeln ('value using poweri is ', x:10:4)
   END
END .
```

For the data

        1214.0
       −0.0021
          1.0
          0.0

the output is

```
1214.0000 =  0.1214 * 10^4
value using poweri is  1214.0000

  -0.0021 = -0.2100 * 10^-2
value using poweri is    -0.0021

   1.0000 =  0.1000 * 10^1
value using poweri is     1.0000

   0.0000 =  0.0000 * 10^0
value using poweri is     0.0000
```

Notes

(i)   The headings of the FUNCTION declarations in the calling program correspond exactly to the mathlib specifications given in appendix B. This is important because, when the compiled program is linked with the library, there is usually no type-checking between the formal parameter lists in the calling program and in the actual library routines.

(ii)   The EXTERN statements tell the compiler that the code of the functions is external to the program. (The use of EXTERN in place of the subprogram body is an extension to standard Pascal.)

(iii)   The operating system must be informed which library is being used, that is, where to find the external code, when the program is loaded. The details are implementation-dependent; readers should consult the documentation for their computer systems.

(iv)   The parameters x in sign and x, i in poweri are formal parameters. When the functions are invoked, any variable names, constants or expressions may be used as actual parameters, provided the types are compatible. □

1.5 FURTHER DATA TYPES

An important feature of Pascal is the variety of data types, both standard and user-defined, which are available in the language. In this section we consider some useful data types which have not been discussed so far.

Characters and Strings

The set of characters available for use in programs and data (for example, the ASCII character set) includes letters, digits, blanks, punctuation marks and basic mathematical symbols. In Pascal we can define constants and declare variables to be of type char, and such quantities then represent single characters.

```
CONST
    space = ' ' ;    (* a constant of type char is a single
                        character enclosed in quotes *)

VAR
    ch  :  char ;    (* ch is a variable of type char *)
```

We can also employ string constants and variables capable of holding a specified number ($\geqslant 1$) of characters, for example

routine: = 'poweri$_{nn}$'

In this case the variable routine would be of type PACKED ARRAY [1..8] OF char. The reader should consult a textbook on Pascal for further details.

Arrays

In solving mathematical problems we frequently require to perform operations on vectors and matrices. To represent a vector or a matrix in Pascal we use an ARRAY, normally of real numbers. We illustrate below how to declare arrays to hold vectors and square matrices of order n. We use a procedure to compute the product of a matrix and a vector.

Example 1.6

```
PROGRAM matrixexample ( input, output ) ;

CONST
    n = 2 ;  (* order of the matrix *)

TYPE
    (* data types for vectors and matrices of order n *)
    vector = ARRAY [1..n] OF real ;
    matrix = ARRAY [1..n, 1..n] OF real ;

VAR
    x, y : vector ;
    A    : matrix ;
    i, j : integer ;


PROCEDURE matmult ( VAR A    : matrix ;
                    VAR x, y : vector  ) ;
(* This procedure computes the product of matrix A and
    vector x, and returns the result in vector y *)

    VAR
        sum  : real ;
        i, j : integer ;
        (* i,j are local variables in the procedure and
            are different from i,j in the main program *)
```

```
BEGIN
(* Form the product Ax and assign the result to y.
   This requires a 'nested FOR loop' *)
FOR i := 1 TO n DO
   BEGIN
   sum := 0.0 ;
   FOR j := 1 TO n DO
      sum := sum + A[i,j]*x[j] ;
   y[i] := sum
   END
END (* matmult *) ;

BEGIN   (* main program *)
FOR i := 1 TO n DO
   FOR j := 1 TO n DO
      read (A[i,j]) ;    (* input the matrix by rows *)
FOR i := 1 TO n DO
   read (x[i]) ;
matmult (A, x, y) ;      (* compute y = Ax *)
FOR i := 1 TO n DO
   writeln (y[i])        (* output the result *)
END .
```

Suitable data would be

1.0 2.0
3.0 4.0
1.0 2.0

The result is

```
5.000000000000E+00
1.100000000000E+01
```

that is the matrix product

$$\begin{bmatrix} 1 & 2 \\ 3 & 4 \end{bmatrix} \begin{bmatrix} 1 \\ 2 \end{bmatrix} = \begin{bmatrix} 5 \\ 11 \end{bmatrix}$$

Notes

(i)   We must specify the types of A, x and y in the parameter list of the
procedure. As Pascal does not allow explicit use of ARRAY in this context, we
define the types vector and matrix globally in the main program. If it were not
for this requirement we could, if we wished, omit the TYPE definition and
declare the arrays as

```
VAR
   x, y : ARRAY [1..n] OF real ;
   A    : ARRAY [1..n, 1..n] OF real ;
```

(ii)   It is necessary to specify the size of the arrays in either a TYPE or a VAR
declaration. The order n is assigned a value in a CONST definition so that it is
easy to locate and easy to change the size of the arrays. (Pascal does not allow
the size of an array to be specified at runtime, for example, by reading in a
value for n.)

(iii)   A and x are variable parameters in PROCEDURE matmult even though they are used only as inputs to the procedure. It would be possible to have these as value parameters, but this is not recommended because of the overheads in setting up and storing local copies of arrays. □

Subrange Types

To index the elements of the arrays in example 1.6 we have used subscript variables i, j of type integer. From the definitions of the arrays, however, we see that the subscripts can only sensibly take on values in a subrange, 1 . . n, of integers. Pascal allows us to define *subrange data types*. In the example above we could define

```
TYPE
    subscript = 1..n ;
    vector    = ARRAY [subscript] OF real ;
    matrix    = ARRAY [subscript, subscript] OF real ;
```

and declare the variables i, j to be of type subscript. Any (erroneous) attempt to set i or j to values lying outside the subrange 1 . . n will then be detected as an error by the Pascal system. As we shall see, the use of subrange types is particularly important in Pascal libraries.

Complex Numbers

In some problems (for example, solving quadratic equations) it is convenient, although not essential, to use complex arithmetic. Since Pascal does not provide a data type for complex numbers, we define our own data type. We use the RECORD facility to define a type complex

```
complex = RECORD
                rl,
                imag : real
          END
```

A variable x of type complex has two parts which correspond to the real and imaginary parts of a complex number. The real part is given by x.rl and the imaginary part by x.imag. Both parts are of type real and can be used like real variables. We can assign the value of x to another complex variable y by the usual assignment statement y: = x. However, procedures are required to perform the basic arithmetic operations. The WITH statement enables us to perform operations on the components of a particular complex variable without the need to prefix the variable name.

```
PROGRAM complexample ( input, output ) ;
(* This program illustrates multiplication and input/output
   of complex numbers *)
```

```
TYPE
   complex = RECORD
                 rl,
                 imag : real
              END ;

VAR
   x, y, z : complex ;

PROCEDURE cmult ( x, y : complex ; VAR cprod : complex ) ;

   BEGIN  (* complex multiplication *)
   WITH cprod DO
      BEGIN
      rl   := x.rl * y.rl - x.imag * y.imag ;
      imag := x.rl * y.imag + x.imag * y.rl
      END
   END (* cmult *) ;

BEGIN  (* main program *)
(* Read two complex numbers, each as a pair of reals *) ;
WITH x DO
   read (rl, imag) ;
WITH y DO
   read (rl, imag) ;
cmult (x, y, z) ;  (* compute z = xy *)
write ('product = ') ;
WITH z DO
   writeln (rl, ' ', imag)
END .
```

1.6 GRAPHICAL OUTPUT: PROCEDURE GRAF

To gain insight into numerical computations it is often very useful to be able to plot or sketch the graph of a function, and it is helpful if this can be done automatically. Most computer installations in Britain have the GINO-F or GHOST graphics packages; these enable users to draw graphs on a variety of devices, for example, terminals with graphics capability or hard-copy graph plotters. The graphics packages are written in FORTRAN 66, but FORTRAN routines can be called from a Pascal program in a way similar to Pascal library routines.

For example, GINO-F has a subroutine GRAF which plots the graph of a set of points $(X[I], Y[I], I = 1, \ldots, NPTS)$. To use this in Pascal we would declare

```
CONST
    (* npts set to a positive integer value *)

TYPE
   vector = ARRAY [1..npts] OF real ;

PROCEDURE graf ( VAR  x, y : vector ;   (* arrays of x,y values
                                           to be plotted *)
                      npts : integer ;  (* number of points
                                           to be plotted *)
                      isc  : integer    (* scale convention,
                                           e.g. 0 = linear scale
                                           on both axes *)
                ) ;

   FORTRAN ;
```

The statement FORTRAN in place of the procedure body performs a similar task to EXTERN in the case of Pascal libraries: it tells the compiler that GRAF is a FORTRAN subroutine in an external file. (This is an extension to standard Pascal — see the note on EXTERN in section 1.4.) Interested readers should consult the documentation of their computer systems for details of accessing and running the graphics packages.

For our purposes it is useful to have a simple Pascal procedure in the mathlib library to print a rough graph on an ordinary (non-graphics) terminal or line-printer. We shall give this procedure the same name as procedure graf and a similar parameter list so that it will be straightforward, if we so wish, to replace it in our programs by the more sophisticated routine.

A CONST definition at the head of the Pascal procedure graf specifies the lengths of the x and y axes, measured in units of printing positions on the terminal or lineprinter. Thus, xaxis and yaxis determine how large the graph will be when it is printed out.

```
xaxis = 50 ;
yaxis = 20 ;
```

Also defined are two local data types

```
xrange = 0..xaxis ;
yrange = 0..yaxis ;
```

The graph itself is represented by means of a two-dimensional array of characters

```
graph  :   ARRAY [xrange, yrange] OF char
```

The array graph will contain a 'picture' of the graph; the array elements, representing points in the xy plane, will be assigned space or non-space characters as appropriate.

We now identify four subproblems:

(1)   Determination of the maximum and minimum values in the arrays x and y;

(2)   Initialisation of the graph array (this involves placing the x and y axes in graph and setting the remainder of the array elements to the space character);

(3)   Representation of the points $(x[l], y[l], i = 1, \ldots, npts)$ in the graph array;

(4)   Output of graph.

We shall develop the solutions to these four subproblems as four separate Pascal procedures: maxmin, inigraph, setgraph and outgraph. Procedure graf will then be constructed as follows (range is a subrange data type in the mathlib library to be defined in chapter 2, and rvector = ARRAY [range] OF real).

```
PROCEDURE graf ( VAR fileout : text ;
                     npts     : range ;
                 VAR x, y     : rvector ) ;
(* This procedure plots the graph of the set of points
   (x[i], y[i], i = 1,...,npts) and outputs the graph
   to the file fileout. The true x axis is printed as a
   row of dots if it lies within the frame of reference *)

   CONST
      xaxis = 50 ;
              (* length of the x axis in printing positions *)
      yaxis = 20 ;
              (* length of the y axis ..    ..        ..    *)

   TYPE
      xrange = 0..xaxis ;  (* range of the graph subscript for
                               x axis *)
      yrange = 0..yaxis ;  (* range of the graph subscript for
                               y axis *)

   VAR
      (* The graph is represented as a 2-dimensional array of
         characters *)
      graph       : ARRAY [xrange, yrange] OF char ;
      xmax, xmin : real ;  (* maximum and minimum values *)
      ymax, ymin : real ;  (* in the arrays x and y      *)


   <PROCEDURE maxmin> ;


   <PROCEDURE inigraph> ;


   <PROCEDURE setgraph> ;


   <PROCEDURE outgraph> ;


   BEGIN  (* body of PROCEDURE graf *)
   (* Call the procedures to plot the graph:                *)
   maxmin (npts, x, xmax, xmin) ;  (* compute xmax, xmin    *)
   maxmin (npts, y, ymax, ymin) ;  (* compute ymax, ymin    *)
   inigraph (ymax, ymin) ;         (* initialise graph array *)
   setgraph (npts, x, y, xmax,     (* represent the points  *)
             xmin, ymax, ymin) ;   (* in the graph array    *)
   outgraph (xmax, xmin,
             ymax, ymin)           (* output the graph      *)
   END (* graf *)
```

The solutions to subproblems (1) and (2) can be coded at once.

```
PROCEDURE maxmin ( n : range ;  VAR a : rvector ;
                   VAR amax, amin : real ) ;
(* Computes the maximum and minimum values in array a[i],
   i = 1,...,n, and returns the values in amax and amin *)

   VAR
      i  : range ;
      ai : real ;

   BEGIN
   amax := a[1] ;  amin := a[1] ;
   FOR i := 2 TO n DO
      BEGIN
      ai := a[i] ;
```

```
    IF ai > amax THEN
        amax := ai
    ELSE
        IF ai < amin THEN
            amin := ai
    END
END (* maxmin *)

PROCEDURE inigraph ( ymax, ymin : real ) ;
(* Initialises the graph array *)

    VAR
        ix : xrange ;   iy : integer ;

    BEGIN
    FOR iy := 1 TO yaxis DO
        BEGIN
        (* Shifted y axis denoted by column of '!' characters *)
        graph[0,iy] := '!' ;
        (* Initialise remainder of graph with ' ' characters *)
        FOR ix := 1 TO xaxis DO
            graph[ix,iy] := ' '
        END ;
    (* Shifted x axis denoted by row of '-' characters *)
    FOR ix := 0 TO xaxis DO
        graph[ix,0] := '-' ;
    (* Insert true x axis if it lies within "frame" *)
    IF ymax <> ymin THEN
        BEGIN
        iy := round ((-ymin/(ymax-ymin))*yaxis) ;
        IF (iy > 0) AND (iy <= yaxis) THEN
            FOR ix := 1 TO xaxis DO
                graph[ix,iy] := '.'
        END
    END (* inigraph *)
```

For subproblem (3) we begin by writing a high-level description and refine this in stages to obtain procedure setgraph.

```
BEGIN  (* subproblem 3 *)
Compute the increments of x and y corresponding to
a single printing position ;
FOR each point (x[i], y[i]) DO
    BEGIN
    determine the closest array element in graph to (x[i], y[i]) ;
    represent the point by placing the character '*' in graph
    END
END
```

The increment in the *x* direction is given by

```
hx := (xmax - xmin)/xaxis
```

and the increment in the *y* direction is obtained similarly (we must not overlook the possibility that hx = 0 or hy = 0). The closest array element to $(x[i], y[i])$ is then found through the predeclared function round, which returns the integer value closest to a given real number

```
ix := round((x[i]-xmin)/hx)
```

and similarly for iy. Finally

```
graph[ix,iy] := '*'
```

The solution to subproblem (3) can now be coded as follows.

```
PROCEDURE setgraph ( npts : range ;   VAR x, y : rvector ;
                         xmax, xmin, ymax, ymin : real ) ;
(* Represents the points (x[i], y[i], i = 1,...,npts)
   as a set of '*' characters in graph array *)

   VAR
      hx, hy  : real ;
      i       : range ;
      ix : xrange ; iy : yrange ;

   BEGIN
   (* Compute the increments of x and y corresponding to
      a single printing position *)
   hx := (xmax - xmin)/xaxis ;   hy := (ymax - ymin)/yaxis ;
   IF hx = 0.0 THEN
      hx := 1.0 ;  (* arbitrary value *)
   IF hy = 0.0 THEN
      hy := 1.0 ;  (* arbitrary value *)
   FOR i := 1 TO npts DO
      BEGIN  (* determine the point in graph array *)
      ix := round((x[i]-xmin)/hx) ;
      iy := round((y[i]-ymin)/hy) ;
      graph[ix,iy] := '*'
      END
   END (* setgraph *)
```

In a similar way we can write a high-level description for subproblem (4) and refine this to obtain procedure outgraph.

```
PROCEDURE outgraph ( xmax, xmin, ymax, ymin : real ) ;
(* Outputs the graph to the file fileout *)

   VAR
      ix : xrange ;  iy : yrange ;

   BEGIN
   writeln (fileout) ;  writeln (fileout) ;
   FOR iy := yaxis DOWNTO 0 DO
      BEGIN
      (* Label the y axis *)
      IF iy = yaxis THEN
         write (fileout, ymax:9, ' ')
      ELSE
         IF iy <> 0 THEN
            write (fileout, '                ')
         ELSE
            write (fileout, ymin:9, ' ') ;
      FOR ix := 0 TO xaxis DO
         write (fileout, graph[ix,iy]) ;  (* output a row
                                             of the graph *)
      writeln (fileout)
      END (* of iy loop *) ;

   (* Label the x axis *)
   write (fileout, '          ', xmin:9) ;
   FOR ix := 1 TO xaxis - 10 DO
      write (fileout, ' ') ;
   writeln (fileout, xmax:9) ;
   writeln (fileout) ;  writeln (fileout)
   END (* outgraph *)
```

We note that the array graph is global to all the subprograms in procedure graf. Since the subprograms perform specific, sequential tasks on graph and employ the array as a 'blackboard', no confusion should result from this. An example of the use of procedure graf is given in section 2.3.

EXERCISES

1.1   The following fragment of Pascal is intended to compute the sum
$\sqrt{0.1} + \sqrt{0.2} + \ldots + \sqrt{10.0}$

```
x := 0.0 ;   sum := 0.0 ;
REPEAT
   x := x + 0.1 ;
   sum := sum + sqrt(x)
UNTIL x = 10.0
```

There is a serious mistake of control in this code. What is the mistake and how should it be corrected?

1.2   Modify program integral (section 1.1) so that the integrand is defined by a Pascal function f(x). Use the program to compute estimates of

$$\int_0^{\pi/2} \sin x \, dx, \quad \int_1^2 x \ln x \, dx, \quad \int_0^2 x^2 e^{-x^2} \, dx$$

with 10, 100 and 1000 steps in each case (only the constants and the body of the function need be altered when the program is applied to each integral). Calculate by hand the error of the three estimates to the first integral. Can you now formulate a conjecture on the accuracy of the mid-point rule as a function of stepwidth?

1.3   Modify procedure quad (section 1.3) so that it will not fail on zero division, that is, the procedure should be able to handle correctly the cases a = 0 and a = b = 0. Use an integer parameter to indicate the number of roots (2, 1 or 0). Test on the equations

$$x^2 + x - 2 = 0, \quad 2x + 3 = 0, \quad 5 = 0$$

Can you find data which will still cause a runtime failure on your computer?
    Generalise quad further so that it can compute complex as well as real roots (use the data type complex discussed in section 1.5). Test the resulting code as thoroughly as you can.

1.4   Write two programs invoking the library functions powerr and poweri to compute and print the values of x^r and x^i, where x, r and i are read in as data. Test on the following:

$$2.5^{-1.8},\ 9.2^{0.0},\ 0.0^{2.6},\ 10.0^{-1000.0},\ (-1.2)^{0.9},$$
$$(-1.1)^{-25},\ (-5.4)^{0},\ 0.0^{10},\ (-0.1)^{1000},\ 0.0^{0}.$$

(A compiler directive may be required at the head of a program which calls a library routine.) Trace the detailed action of the functions in the various cases by referring to section 1.4 and the library codes in appendix C.

1.5 Devise a search method using the library function sign to locate a value of x which satisfies the equation $x^3 - c = 0$, where c is a constant lying between 0 and 1. Hence compute $0.5^{1/3}$ correct to 2 decimal places. Repeat the exercise to find an approximate solution of $x + \ln x = 0$ (you may find procedure graf useful to localise the search). Can you show that this is the only solution of the equation?

1.6 Modify procedure matmult (section 1.5) to obtain
(i) a Pascal function which computes the inner product

$$\sum_{i=1}^{n} x_i y_i \text{ of two } n\text{-component vectors } \mathbf{x} \text{ and } \mathbf{y}$$

(ii) a procedure which computes the matrix product $\mathbf{C} = \mathbf{AB}$ of two $n \times n$ matrices $\mathbf{A}$ and $\mathbf{B}$, where

$$c_{ij} = \sum_{k=1}^{n} a_{ik} b_{kj}$$

Check that the routines perform correctly on a set of vectors and matrices of orders 2 and 3, then try some larger cases.

1.7 Write a procedure cdiv which computes the quotient of two complex numbers and returns the result in a variable cquot of type complex (first make sure that you understand how the quotient of two complex numbers is defined!). Incorporate the procedure in a program similar to program complexample and test on the following

$$i/1,\ 1/i,\ 1/2,\ (2+i)/(1-2i),\ i/0$$

1.8 Develop a program fungraph which uses procedure graf to plot the graph of a function $f(x)$, sending the result to a named file rather than to the output file. (See program polgraph in section 2.3. You will need an extra filename in the program heading and other details — refer to a textbook on Pascal.) Hence prepare on a disk file the graph of $y = \sin(x^2)$ over an interval $a \leqslant x \leqslant b$. A message should also be sent to output stating the interval and tabulation step. Use the program to check if figure 1.1 is correctly drawn.

2 Principles of Mathematical Software

Let me taste your ware.

Anon., *Simple Simon*

We now consider the general principles we should bear in mind when developing mathematical software, with particular reference to the structure and design of the mathlib library.

In developing any piece of software we should be guided by the principles of good programming practice, and much of what is said here is simply the procedure we should aim to follow whenever we write a program. The process of producing mathematical software involves

(i)   Design and analysis of algorithms;
(ii)   Coding the algorithms (software realisation);
(iii)   Extensive testing;
(iv)   Detailed documentation;
(v)   Distribution and maintenance of the software.

Stages (i) and (ii), in particular, are sometimes difficult for the beginner to distinguish. Stage (i) requires us to define precisely the steps we must take to solve the problem (this is the *algorithm*); stage (ii) involves software construction (the actual coding of the algorithm).

2.1   DESIGN AND ANALYSIS OF ALGORITHMS

To illustrate what is meant by an algorithm, we consider the problem of evaluating a polynomial, that is, an expression of the form

$$p_n(x) = a_0 + a_1 x + \ldots + a_n x^n$$

for a specified value of x. The *coefficients* $a_0, a_1, \ldots, a_n$ are constants, and n is an integer called the *degree*. (Polynomials play a central role in numerical computation and will be discussed further in chapter 3.)

To evaluate a polynomial we must first devise an algorithm. The most obvious method is to use a power function.

Algorithm 2.1

```
Input  n, a[0], ..., a[n], x ;

sum := a[0] ;
FOR i := 1 TO n DO
   sum := sum + a[i]*x^i ;

Output sum .
```

Although we have expressed the algorithm partly in Pascal, we could equally
well have employed any clear and convenient notation. Using a notation
suggested by Vandergraft (1978) we would have

    1   Input $\{n, a_0, a_1, \ldots, a_n, x\}$
    2   $S \leftarrow a_0$
    3   For $i = 1, 2, \ldots, n$
       3.1 $S \leftarrow S + a_i x^i$
    4   Output $\{S\}$

Here we are using a different notation (language) to express the same algorithm.
We shall express the algorithms in this book in a Pascal-like language because the
reader is familiar with Pascal and also, of course, because we intend ultimately to
code the algorithms in Pascal. Since we are interested mainly in the mathematical
aspects of the algorithms, we will usually omit input/output statements.

    Having once devised an algorithm, we can study how to improve it (to make it
more comprehensive, accurate or efficient). A little reflection shows that it is
inefficient to use a power function in algorithm 2.1, because we have to evaluate
each of the powers $x^1, x^2, \ldots, x^n$ separately. We would do better to save the
partial power as follows.

Algorithm 2.2

```
sum := a[0] ;   xi := x ;
FOR i := 1 TO n DO
   BEGIN
   sum := sum + a[i]*xi ;
   xi  := xi*x
   END
```

Here we 'remember' the value of x^i from one iteration to the next to avoid
recomputing it.

    The idea of first formulating a basic working algorithm and then improving it
in stages is a technique of considerable importance. This approach can sometimes be used to establish the correctness of the final algorithm.

    An algorithm for polynomial evaluation based on a rather less obvious
approach can be obtained by writing

$$p_n(x) = (\ldots ((a_n x + a_{n-1})x + a_{n-2})x + \ldots + a_1) x + a_0$$

This is the so-called *nested* form; the highest coefficient a_n appears within the innermost parentheses. If we now define

$$b_{n-1} = a_n$$
$$b_{n-2} = b_{n-1}x + a_{n-1}$$
$$b_{n-3} = b_{n-2}x + a_{n-2}$$

$$\cdot \qquad \cdot \qquad \cdot$$
$$\cdot \qquad \cdot \qquad \cdot$$

$$b_{-1} = b_0 x \quad + a_0$$

we see that b_{-1} takes the value of $p_n(x)$. This method of evaluating a polynomial is known as *Horner's scheme*. We shall see in chapter 3 that the intermediate values $b_{n-1}, \ldots, b_0$ can be used to derive further information about the polynomial; for the moment, however, we are interested only in the final value b_{-1}. We summarise the method as

Algorithm 2.3

```
(* Horner's scheme *)
b := a[n] ;
FOR i := n-1 DOWNTO 0 DO
   b := b*x + a[i] ;
(* The final value of b is the value of the polynomial *)
```

We can test algorithm 2.3 by 'hand running' it for particular cases. Careful hand running of an algorithm or a program on representative data is a good way of discovering faults.

Example 2.1

Evaluate the cubic

$$p(x) = x^3 - 6x^2 + 11x - 4$$

at $x = 1$.
   We have

$$n = 3, \ x = 1$$

and the coefficients a[n], ..., a[0] are

$$1, -6, 11, -4$$

Following algorithm 2.3

```
b = 1;
i = 2; b = 1 × 1 − 6 = −5;
i = 1; b = −5 × 1 + 11 = 6;
i = 0; b = 6 × 1 − 4 = 2
```

We conclude that the value of the cubic at $x = 1$ is 2 (this is correct; we can check the result by algorithm 2.1). □

Algorithm 2.3 solves correctly the same class of problems (evaluation of polynomials) as algorithms 2.1 and 2.2, but it is more efficient. For a polynomial of degree n we can show (by counting the operations involved in each iteration) that the number of arithmetic operations required to compute the answer is

	Multiplications	Additions
Algorithm 2.1	$n(n + 1)/2$	n
Algorithm 2.2	$2n$	n
Algorithm 2.3	n	n

Clearly, algorithm 2.3 is the most efficient of the three; whether it is of optimal efficiency is a more difficult question which leads into computational complexity theory.

We conclude this section by listing the main properties to look for in an algorithm which is to be implemented in a mathematical library. (It is important to bear in mind that general users of the library will have only a hazy idea of how the library routines work, that is, the routines will appear to them as 'black boxes'.)

Stability The algorithm must not be unduly sensitive to rounding errors, which are unavoidable in computing with real numbers. An algorithm with this essential property, such as Horner's scheme for polynomial evaluation, is said to be *stable*. Examples of stable and unstable algorithms will be given in chapter 4.

Robustness The algorithm should solve correctly the problems for which it is intended. In cases where it cannot obtain a solution, the algorithm should terminate with a sensible error message.

Accuracy The algorithm should be able to solve problems to an accuracy predicted in the documentation, that is, it should be reliable.

Efficiency When implemented on a computer, the algorithm should not take an excessive amount of time or store to solve problems. One measure of efficiency (considered above for polynomial evaluation) is the number of arithmetic operations required to compute a solution.

2.2 SOFTWARE REALISATION: THE MATHLIB LIBRARY

In earlier days of computing it was the tacit aim of programmers to employ clever tricks in order to utilise fully the meagre computing resources then available. Thus we find in the library of an early computer called EDSAC the following description of a routine to solve differential equations:

...G1 should be placed in the upper half of the store to obtain maximum accuracy (the ideal position is 386 onwards). This is because one of the orders forms part of a constant which thus depends slightly on the location of the routine. In normal use the effect is quite negligible...
(quoted by W. R. Cowell and L. D. Fosdick in Rice (1977))

Today, rapid advances in computer technology mean that on large, mainframe computers the scarce resource is not computing power but programming power. Considerable effort is required to produce good mathematical software, and hence the emphasis is on being able to transport well-tried programs from one installation to another with a minimum of modification. This means, in particular, that programmers should resist the temptation to exploit peculiarities of the computer system or obscure features of the programming language; tricks of this sort make programs difficult to understand and may also make them fail when implemented on different systems.

We should design our algorithms with the aim of *adaptability* and code them with the idea of transporting the software from one computer to another. For example, a test of the form

```
IF abs(error) < 0.5E-10 THEN ...
```

is not adaptable because (due to the different 'precision' of computers − see chapter 4) it may not be possible to obtain the required accuracy on some machines. The algorithm will therefore behave differently on different systems and may fail on some. Error tests, as we shall see repeatedly throughout this book, must be expressed relative to the precision of the computer. Similarly, rational numbers which do not have a finite decimal representation, such as $1/3$, should be written in their rational form $(1/3)$ in a program.

The algorithms in this book will be coded as subprograms and collected together in the mathlib library. A library in Pascal is a program. A basic convention of Pascal is that all programs have the same underlying structure; thus the general structure of a Pascal library is

```
PROGRAM <library name> ( input, output ) ;

CONST
   <definition of library constants, e.g. array bounds> ;

TYPE
   <definition of library data types, e.g. array types> ;

<FUNCTION and PROCEDURE declarations
 - the actual code of the library routines> ;

BEGIN   (* dummy main program *)
END .
```

We note that the main program body is empty (the program does nothing if executed!). The library is compiled in the normal way except that it is necessary to instruct the compiler to save the entry point names, that is, the names of the subprograms; these are required when the library is linked to a user program.

Generally the simplest way to effect this instruction is to insert the compiler directive

```
(*$E+ *)
```

as the first line of the library above the program heading.

A word of warning is necessary. Libraries are not defined in standard Pascal; the exact form of a Pascal library is therefore implementation-dependent and may differ from the above in some details, particularly in the program heading (this is discussed further in appendix D).

Conventions of the mathlib Library

We have stressed the importance of making algorithms adaptable and coding them as portable software. To make the mathlib library portable we define the machine-dependent constants in a CONST definition at the head of the library. This may be regarded as an extension to the predefined constants of Pascal (for example, maxint, the largest integer). The following constants will be used (the values are those appropriate to the Amdahl V/7 using REAL*8; they should be modified for different computer systems).

```
smallreal = 1.0E-74 ;   (* 100 times the smallest positive
                            real number *)
rprec4    = 0.9E-15 ;   (* 4 times the relative precision
                            of the computer *)
smallog   = -170.0  ;   (* ln(smallreal), used in tests to
                            avoid underflow failure *)
```

The 'smallest positive real number' is defined as the smallest real number *rmin* such that both *rmin* and *−rmin* are representable as non-zero on the computer; the factor of 100 is chosen arbitrarily to ensure that smallreal is substantially larger than *rmin*. The 'relative precision' enables us to calculate the maximum possible rounding error incurred in representing real numbers on the computer; rprec4 can then be used to express error tests in an adaptable form. Further discussion and programs to determine the machine constants are given in chapter 4.

We also define in a TYPE definition the special data types used by the routines of the library. These include the array subscript ranges and array types

```
(* To represent real and integer vectors *)
range    = 1 .. upbnd ;  (* range of the array subscript *)
rvector = ARRAY [range] OF real ;
ivector = ARRAY [range] OF integer ;

(* To represent square matrices of real numbers *)
matrix  = ARRAY [range, range] OF real ;

(* To represent the coefficients of a polynomial *)
index = 0 .. upbnd ;  (* range of the array subscript *)
coeff = ARRAY [index] OF real ;
```

A serious weakness of current standard Pascal is that all arrays, including arrays used in subprograms, must be of constant size. It is proposed to extend Pascal to allow arrays of variable size in subprograms (Addyman (1980)). As this extension is not yet widely available, we use fixed-size arrays and define the constant upbnd at the head of the library (upbnd is currently set to 100; this can be altered if required). We have designed our conventions so that it should be straightforward to modify the library when the extension becomes available. The necessary changes to the library are discussed in section 2.4.

A disadvantage of library routines from the user's point of view is the need to supply a long and sometimes confusing parameter list. Library routines are often judged on how painless they are to use rather than how 'clever' they may be. The design of the parameter list is therefore of considerable importance. We can assist the user by keeping the number of parameters to a minimum, ideally only those necessary to specify the problem. We also adopt a consistent sequence based on the NAG library model. The order of the parameters (any or all of which may be absent) is *filenames, subprogram names, input parameters, input/ output parameters, output parameters* and *error indicator*. All but the scalar input parameters and the subprogram names will be preceded by VAR (arrays, even as inputs, will be preceded by VAR as explained in the notes on example 1.6, chapter 1). To avoid side-effects, only input parameters will be used in FUNCTION subprograms.

Error Messages

The *error indicator* is an integer variable which signals whether a computation has been performed successfully or not; if not, it returns an indication of the nature and severity of the error. For example, in a routine to solve non-linear equations it may not be possible to compute a solution to the accuracy requested by the user; in this case the routine will return the best answer it can and set the error indicator to warn the user that the required accuracy has not been attained.

A library utility routine is provided in the mathlib library to handle errors and non-standard conditions in a consistent way. For this purpose three additional data types are defined in the library

```
(* For transmitting error messages *)
name     = PACKED ARRAY [1..8] OF char ;
message  = PACKED ARRAY [1..30] OF char ;
severity = (fatal, warning) ;
```

The error-handling routine is

```
PROCEDURE errormessage ( routine      : name ;
                         errornum     : integer ;
                         errorsummary : message ;
                         condition    : severity ;
                         VAR eflag    : integer ) ;
(* This procedure handles error messages and warnings in the
   mathlib library. A message is written to the output file if
```

```
eflag=0 on entry. If condition=fatal, an error message is
issued; if condition=warning, a warning is issued.
A message consists of the name of the routine (routine),
the error number (errornum) and a summary of the error or
warning (errorsummary). The definition of the library data
types name and message means that routine and errorsummary
must be strings of length 8 and 30 characters respectively.
In all cases the error indicator eflag is returned with
the value of errornum *)

BEGIN
IF eflag = 0 THEN
    CASE condition OF
        fatal:
            writeln ('*** error in ', routine, ':', errornum:2,
                     '  ', errorsummary) ;
        warning:
            writeln ('*** warning in ', routine, ':', errornum:2,
                     '  ', errorsummary)
        END (* CASE *) ;
eflag := errornum
END (* errormessage *)
```

Procedure **errormessage** is called by many of the library routines when
error conditions are encountered. Two classes of error are distinguished:

> condition = fatal: the library routine has failed to compute a sensible
> solution
> condition = warning: the library routine has been able to compute a
> solution but not to the required accuracy

Procedure **errormessage**, when called by a library routine with **eflag** = 0,
writes an appropriate message to the output file and returns eflag set to the value
of errornum. The significance of the error numbers is detailed in the specification
of each library routine. A typical use of errormessage is found in the library procedure polnewt (see chapter 5)

```
errormessage ('polnewt ', 1,
    'iteration fails to converge   ', fatal, eflag)
```

Further information on error handling is given in the summary of the library in
appendix B.

It is sometimes inconvenient to have a message written out in the form trans-
mitted by errormessage. The library routine may be part of a larger computation
for which a more useful recovery action can be taken. To allow for this, the
output of error messages and warnings is suppressed if the routine is called with
eflag $\neq$ 0. It is then the responsibility of the user to check the error indicator in
the calling program on return from the library routine (this should be normal
practice in any case).

We now give a simple example of a library routine written according to the
mathlib conventions. This is a FUNCTION subprogram to evaluate a polynomial
by the method of algorithm 2.3.

Example 2.2

```
FUNCTION polsum (      n : index ;
                  VAR  a : coeff ;
                       x : real    ) :  real ;
(* This function evaluates the polynomial
        a[0] + a[1]*x + ... + a[n]*x^n
   for a specified value of x and n >= 0 using Horner's scheme *)

   VAR
      b : real ;
      i : index ;

   BEGIN
   b := a[n] ;
   FOR i := n-1 DOWNTO 0 DO
      b := b*x + a[i] ;
   polsum := b
   END (* polsum *)
```

Notes

(i)   An array of type coeff is used to pass the values of the coefficients to the subprogram.

(ii)   The degree n is restricted to the range 0 . . 100 defined by the subrange type index. The Pascal system will detect an error if we attempt to set n to a value lying outside this range; hence it is superfluous to check for inconsistent data (for example, n < 0) in polsum. This is a convenient, general way of ensuring that array bounds are not violated. □

An example program using function polsum and procedure graf from the library is given at the end of the next section.

2.3   NUMERICAL TESTING AND DOCUMENTATION

The numerical testing and assessment of mathematical routines is a difficult and time-consuming task which has received much attention in recent years. There are two main objectives in testing a routine:

(i)   to establish experimentally the correctness of the code;
(ii)   to measure the relative performance of the routine against other routines.

In order to establish the correctness of the code, we select a set of test problems known as *code exercisers* which are designed to test as many of the main paths as possible (including the error paths) of the program. (There now exist 'branch analyser' programs which verify if a set of problems does indeed use the main paths.) If possible, it is advisable to hand run the code and check that the hand-run results agree with results obtained by executing the program; this is especially useful for locating faults.

Having ensured as far as possible that the routine is correctly coded, we require to measure its performance. It is important to understand that we are evaluating, not the numerical method in abstract, or even the algorithm, but the code realisation of the algorithm. Different programmers can write for the same numerical method programs which perform very differently.

Nearly all performance evaluation is carried out by the *battery method*: a large set of problems which the routine has been designed to solve is prepared, and the performance of the routine is measured against that of existing routines. We would record, for example, the amount of central processor time required to solve each problem, whether it has been solved correctly and, if so, how accurately. Unfortunately, it is often difficult, even after extensive testing, to say whether one routine is absolutely better than another. We find typically that one routine performs well on some problems and less well on others, while the opposite may be the case for another routine. Hence in a library we usually find a selection of routines to solve the same class of problem, for example, numerical evaluation of definite integrals. Where appropriate we shall use a form of the battery method to assess the performance of the routines in this book.

Documentation of Library Routines

The documentation of programs and particularly of library routines is of major importance. From the user's point of view the documentation for a routine should give

(i)   A clear statement of the class of problem which the routine is designed to solve, any cases it is particularly suitable for and any restrictions on its applicability;
(ii) A precise statement of the type and purpose of each parameter in the routine, together with any restrictions on the parameters;
(iii) An example program to illustrate the use of the routine (users will often run the example program and then adapt it to the particular problem they wish to solve);
(iv) Some discussion of the accuracy, speed and storage requirements of the routine, together with references for further study.

It is also important to spare some thought for implementers wishing to mount the library on different computer systems or to modify the code. Any non-standard features used in the code must be pointed out, and the main sections of the programs should be highlighted by the use of comments and clear structure. Even though the general user may never see the code, it is important to bear in mind that someone has to maintain and extend it; this is far easier if the software is well structured and carefully documented.

Detailed examples of user documentation will be found in the manuals of a commercial library such as the NAG (Numerical Algorithms Group) library; the reader is advised to consult these for guidelines on design and layout of documentation. Space does not permit us to give full user specifications in this book. As far as we are concerned, points (i) and (ii) above are covered in the summary given in appendix B, while points (iii) and (iv) will be developed for each routine in the appropriate chapter.

An Example Program

To illustrate how we would use the library, we consider the problem of plotting the graph of the polynomial

$$p(x) = 1 - 32x + 160x^2 - 256x^3 + 128x^4$$

over $0 \leqslant x \leqslant 1$. We use two routines, function polsum and procedure graf, from the library. In appendix B we find the descriptions of these routines.

```
PROCEDURE graf ( VAR fileout : text ;   npts : range ;
                     VAR x, y     : rvector ) ;
Plots the graph of the points (x[i], y[i], i = 1,...,npts),
where npts <= upbnd. Graph is written to the file fileout.
[1.6, 2.3]


FUNCTION polsum ( n : index ;   VAR a : coeff ;   x : real )
               : real ;
Evaluates the polynomial of degree n >= 0,
   a[0] + a[1]*x + ... + a[n]*x^n,
for a specified value of x using Horner's scheme.
[2.2, 2.3]
```

References in square brackets are to the sections of the book where discussion and codes of the library routines can be found.

To plot the curve $y = p(x)$ we first compute the y values ($y[i]$, $i = 1, \ldots,$ npts) at a suitable set of points ($x[i]$, $i = 1, \ldots,$ npts). We then supply the arrays x and y to procedure graf. The first parameter of graf is a filename which allows the user to direct where the graph is to be plotted. Thus, if we wish the graph to appear on the output file, the actual parameter will be output, and we invoke graf as

```
graf (output, npts, x, y)
```

The example program and results follow.

```
PROGRAM polgraph ( input, output ) ;
(* This program uses two mathlib library routines:
      polsum  to evaluate a polynomial,
      graf    to plot a graph of the polynomial *)

CONST
   npts =    50 ;   (* number of graph points *)
   aa   =   0.0 ;   (* interval over which the polynomial *)
   bb   =   1.0 ;   (* is to be plotted                   *)
   n    =     4 ;   (* degree of the polynomial *)

(* Define data types corresponding to those used
   in the mathlib routines *)
TYPE
   index   = 0..100 ;
   range   = 1..100 ;
   coeff   = ARRAY [index] OF real ;
   rvector = ARRAY [range] OF real ;
```

```
VAR
   xx, hx : real     ;
   i      : range    ;
   a      : coeff    ;
   x, y   : rvector  ;

(* Declare the library routines used in the program *)
FUNCTION polsum (      n : index ;
                  VAR a : coeff ;
                      x : real     ) : real ;
   EXTERN ;

PROCEDURE graf ( VAR fileout : text ;
                     npts    : range ;
                 VAR x, y    : rvector ) ;
   EXTERN ;

BEGIN
(* A graph of the polynomial
     a[0] + a[1]*x + a[2]*x^2 + a[3]*x^3 + a[4]*x^4
   is plotted over the interval aa <= x <= bb  *)

(* Specify the coefficients of the polynomial *)
a[0] := 1.0 ;  a[1] := -32.0 ;  a[2] := 160.0 ;
a[3] := -256.0 ;  a[4] := 128.0 ;

(* Use polsum to evaluate the polynomial at a
   set of equally-spaced points from aa to bb *)
hx := (bb - aa)/(npts - 1) ;
FOR i := 1 TO npts DO
   BEGIN
   xx := aa + (i-1)*hx ;
   x[i] := xx ;               (* store the x value *)
   y[i] := polsum (n, a, xx)  (* compute and store the y value *)
   END ;
writeln ;  writeln ;          (* print the heading *)
writeln ('                      graph of polynomial') ;
writeln ;
graf (output, npts, x, y)     (* plot the graph *)
END .
```

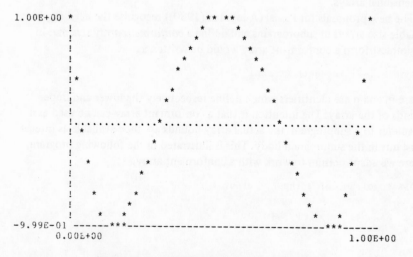

Figure 2.1   Output from program polgraph

Notes

(i)   This is a typical example program which might appear in the documentation of polsum. For the sake of clarity and convenience in examples, it is quite
usual to specify the data by constant definitions and assignment statements
rather than by reading the data from a file. We have done this for the degree n
and the coefficients a[0] , . . ., a[4] of the polynomial. To plot a different
polynomial the appropriate lines of the program would be altered.

(ii)   The headings of the routines polsum and graf appear in the example program exactly as they are declared in mathlib. The same subrange types, index and
range, are also used. This is a worthwhile practice, since on most systems there is
unlikely to be any type-checking between the routines declared in the user program and the actual library codes.

(iii)   Even though the program does not read input data, we still include the
input file in the program heading. This is to make the heading of polgraph
compatible with that of mathlib.

Program polgraph can be modified to send the graph to a named file
rather than to the output file. The details of this are left as an exercise for the
reader.

2.4   VARIABLE-SIZE (CONFORMANT) ARRAYS

The requirement of constant sizes for arrays in subprograms is a severe limitation on libraries of mathematical routines. It is wasteful of storage to use an
array capable of holding a polynomial of degree 100 if we require to evaluate
only low-degree polynomials; on the other hand it is impossible, without redefining the array bound in the library, to evaluate polynomials of degree
greater than 100. The same objections apply with added force to multi-
dimensional arrays.

The new proposals for Pascal (Addyman (1980)) recognise the need for
variable-size arrays in subprograms and define a *conformant array schema*. In
its simplest form a conformant array would be written as

```
ARRAY [m..n : integer] OF real
```

where m and n are identifiers which define respectively the lower and upper
bounds of the array. The intention is that a conformant array can be used as a
parameter to a subprogram; the actual array bounds are then available as integer
constants in the subprogram body. This is illustrated in the following program,
where we adapt polsum to work with a conformant array.

```
PROGRAM polsumconf ( input, output ) ;

VAR
    a : ARRAY [0..2] OF real ;
    x : real ;
```

```
FUNCTION polsumc ( VAR a : ARRAY [zero..n : integer] OF real ;
                       x : real ) :  real ;

    VAR
       b : real ;
       i : integer ;

    BEGIN
    <body of function>
    END (* polsumc *) ;
    .
BEGIN
(* Compute and print the value of the polynomial
                1 + 2*x + 3*x^2                            *)
a[0] := 1.0 ;   a[1] := 2.0 ;   a[2] := 3.0 ;
read (x) ;
writeln (polsumc(a, x))
END .
```

We note the use of zero (not 0) in the subprogram heading to comply with the
conformant schema. We also note that it is no longer necessary to supply the
degree of the polynomial as an actual parameter when the subprogram is invoked
(the value n = 2 is supplied through the conformant schema).

Although the new proposals for Pascal do not include the definition and use
of libraries, it is likely that compilers which implement these proposals will
allow the use of conformant arrays in Pascal libraries (indeed this is likely to be
one of the main uses!). The mathlib library requires minor changes to accommodate conformant arrays.

(i)   Delete the definitions of array types (for example, coeff) and ranges (for
example, index) from the library;
(ii)   Replace the formal array parameters in function and procedure headings by
the appropriate conformant arrays, and remove any parameters concerned with
array sizes;
(iii) Remove any array size parameters from the actual parameter lists in cases
where library routines are invoked from other library routines.

For example, the heading of graf would become

```
PROCEDURE graf ( VAR fileout : text ;
                 VAR x, y    : ARRAY [one..npts : integer]
                                                 OF real )
```

Modifications on the lines of (iii) will also be required in user programs which
invoke routines from the library.

It should be stressed that the facility for variable-size arrays will be an extension, not a replacement, of existing Pascal capabilities; the mathlib library will
still be usable in its unmodified form.

Further Reading

The production and evaluation of mathematical software is a subject which has
been developing rapidly in recent years. An interesting survey article is given by

W. R. Cowell and L. D. Fosdick in *Mathematical Software* (Rice (1977)). Many modern FORTRAN codes for mathematical computations are given in the very useful book *Computer Methods for Mathematical Computations* (Forsythe, Malcolm and Moler (1979)). This book can be recommended to the reader who has a basic knowledge of FORTRAN 66.

The design, production and documentation of a large modern mathematical library is discussed in a paper by J. K. Reid and M. J. Hopper in the conference proceedings *Production and Assessment of Numerical Software* (Hennell and Delves (1980)). These proceedings also contain a number of useful contributions covering software verification, portability and general design of a mathematical library.

EXERCISES

2.1 Evaluate the polynomial

$$p(x) = 2x^3 + 5x^2 - 4$$

at $x = -2$ by 'hand running' FUNCTION polsum. Can you now solve the equation $p(x) = 0$?

Write a program using polsum to tabulate a polynomial over an interval $a \leqslant x \leqslant b$ in steps of h (a, b and h read in as data). Take $a = -2.0$, $b = 1.0$, $h = 0.1$, and run the program on $p(x)$ above.

2.2 By counting the number of arithmetic operations in each iteration of algorithms 2.1, 2.2 and 2.3 and summing as appropriate, show that the total number of operations is as given in the table in section 2.1. (Assume that, for example, x^3 is evaluated as $x.x.x$).

If you had to evaluate a polynomial of degree 100 using programs based on the three algorithms, what would you expect to find for the relative computation times? (Assume that multiplication takes the same time as addition). Would you expect the numerical results to be the same?

2.3 Devise two efficient schemes modelled on algorithms 2.2 and 2.3 to evaluate the polynomial

$$p_n(x) = 1 + x + \frac{x^2}{2!} + \ldots + \frac{x^n}{n!}$$

(You should not use an array to store the coefficients!) Is there any reason to prefer one scheme rather than the other?

2.4 Modify program polgraph to plot the graph of the cubic

$$p(x) = 2x^3 - 4x^2 - 3x + 4$$

over $-1.5 \leqslant x \leqslant 2.5$. By taking successively smaller intervals, locate the root of $p(x) = 0$ which lies closest to 0.0, correct to 1 decimal place. Are there any other real roots of this equation?

2.5   The following approximations to $\log_{10}x$ and 10^y are accurate to five decimal places.

(a) $\log_{10}(x) \approx 0.5 + c_1B + c_3B^3 + c_5B^5 + c_7B^7$

where

   $B = (x - \sqrt{10})/(x + \sqrt{10})$

and

   $c_1 = 0.86855434, \; c_3 = 0.29115068,$
   $c_5 = 0.15361371, \; c_7 = 0.21139497$

(provided that $1 \leqslant x \leqslant 10$)

(b) $10^y \approx (1 + a_1y + a_2y^2 + a_3y^3 + a_4y^4 + a_5y^5)^2$

where

   $a_1 = 1.15138424, \; a_2 = 0.66130851, \; a_3 = 0.26130650,$
   $a_4 = 0.05890681, \; a_5 = 0.02936622$

(provided that $0 \leqslant y \leqslant 1$).

Write a function power which uses these formulae to compute x^r for given x and r, provided $1 \leqslant x \leqslant 10$ and $0 \leqslant rx \leqslant 6$. In other cases the function should return an error indicator set to 1. The function should make use of the routines polsum and errormessage from the mathlib library.

2.6   Extend function power in exercise 2.5 to compute x^r for any values of x and r for which the result is mathematically defined (see section 1.4). You may use the mathlib routines sign and poweri and any predeclared Pascal functions except exp and ln. (*Hint:* express x in the form $x = f \times 10^m$, where $1 \leqslant f < 10$ and m is an integer, using procedure float from example 1.5.)

   Incorporate the function in a test program and check the function and its error paths, comparing the results with those obtained from the mathlib routine powerr.

3 Basic Mathematics and Applications

Roll up that map; it will not be wanted these ten years.

William Pitt The Younger

When developing software for mathematical computations, we meet, apart from questions of code verification and empirical testing, a number of tasks at a higher level of abstraction. These include

(i)   Devising or selecting suitable algorithms to solve the problems, and being able to understand how these work;

(ii)   Selecting the 'best' algorithms according to certain criteria of efficiency, generality etc.;

(iii)   Proving that the chosen algorithms will work correctly for the class of problem which they are intended to solve.

To approach tasks of this sort we make use of the tools available in mathematics. The purpose of the present chapter is to highlight the mathematical ideas which will be used throughout the book. We shall not lose sight of the point that our main interest is in numerical computation, that is, in producing answers as well as analysing questions. The mathematical development will be reinforced with some of the important numerical methods which follow naturally from classical analysis. These tools and methods will form the basis for developments in the remainder of the book.

3.1   SETS AND FUNCTIONS

The fundamental notion on which all constructs in mathematics are built is that of a *set* of objects. A set is specified by listing the objects (*elements*) either explicitly or implicitly, for example,

$$X_{op} = \{+, -, *, / \}$$
$$I_+ = \{1, 2, 3, \dots \}$$

or by defining the class of objects which is to comprise elements of the set,

$$X_{op} = \{x \mid x \text{ is a real arithmetic operator in Pascal} \}$$
$$I_+ = \{i \mid i \text{ is an integer}, i \geqslant 1 \}$$

To indicate set membership we use the symbol $\in$; thus

$$* \in X_{op} \text{ but } \uparrow \notin X_{op}$$

In numerical computation we are interested in sets which consist of numbers, vectors, matrices or functions. Of particular importance are sets of real numbers.

Definition 3.1

(i) If a and b are two real numbers and $a \leqslant b$, then the set $\{x \mid x$ is a real number, $a \leqslant x \leqslant b\}$ is called a *closed interval*, denoted by $[a, b]$.
(ii) If $a < b$ in (i), then the set $\{x \mid a < x < b\}$ is called an *open interval*, denoted by (a, b).
(iii) The set of all real numbers is denoted by $R = (-\infty, \infty)$. □

As an example,

$$\pi \in R, \quad \pi \in (3.1, 3.2) \text{ but } \pi \notin [3.142, 3.143].$$

If X and Y are two sets and each element of X is also an element of Y, then we say that X is a *subset* of Y and write $X \subseteq Y$. If also $Y \subseteq X$, then the two sets are *equal*; otherwise we write $X \subset Y$ (X is a *proper subset* of Y). Thus,

$$[3.142, 3.143] \subset (3.1, 3.2) \subset R$$

The set R of real numbers is *infinite* and *non-denumerable*; given any two real numbers a and b with $a < b$, we can find another real number x such that $a < x < b$. On the other hand, because the range of real numbers on a computer is finite and the number of significant digits representable in a computer word is finite, the set of real numbers available on any computing machine is *finite*. Denoting the set of 'machine numbers' by R_m, we have

$$R_m \subset R$$

We may devise and to a certain extent analyse our algorithms on the assumption that the full set R of real numbers is available; at some point, however, we are forced to consider the consequences of using the finite set R_m instead of R. The consequences are non-trivial; this fundamental fact of numerical computation will be investigated in chapter 4.

Functions

In order to develop a structure on the basis of sets, we consider a correspondence between the elements of two sets.

Definition 3.2

If X and Y are two sets and a correspondence is established involving pairs of elements, such that to each $x \in X$ there corresponds exactly one $y \in Y$, then the correspondence is called a *function* or *mapping* from X into Y. □

We note that it is not necessary for every element $y \in Y$ to be involved in this correspondence. However, for each $x \in X$ there must be a unique element $y \in Y$ corresponding to it. The correspondence is determined according to some rule (f, say), and we introduce a notation to reflect this: we write $y = f(x)$ and call y 'the value of f at x'. The function or mapping as a whole is written as $f : X \rightarrow Y$.

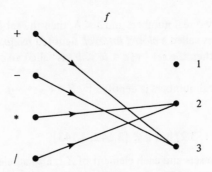

Figure 3.1   Diagram of a function (mapping) f from $\{+, -, *, /\}$
into $\{1, 2, 3\}$: $f(+) = f(-) = 3, f(*) = f(/) = 2$

Example 3.1

Computers represent text characters by mapping them to a subset of positive integers. The details of the mapping depend on the particular character set C_p in use; for the ASCII character set the *range* of the mapping is $N_p = \{32, \ldots, 126\}$. Each element of C_p corresponds to exactly one element of N_p and conversely (this is described mathematically as a *1-1 correspondence* or *bijection*). Thus it makes sense to ask if one character 'precedes' another (in Pascal, for example, the boolean expression 'A' < 'Z' is true).

There are two predeclared Pascal functions concerned with this mapping and available to the programmer. They are

(i)   ord : $C_p \rightarrow N_p$

For example

ord ('A') = 65, ord ('Z') = 90.

(ii)   chr : $N_p \rightarrow C_p$

For example

chr (65) = 'A', chr (90) = 'Z'   (ASCII conventions).

For obvious reasons, ord and chr are said to be *inverses* of each other; in mathematical notation we would write

chr = ord^{-1}

meaning that if ch is any argument of type char, then

    chr (ord (ch))

has the same value as ch. □

The functions we shall be most concerned with, and those most familiar to the reader, are of the form

$$f : X \to R, \quad X \subseteq R$$

that is, *real-valued* functions of *one real variable*. These map the points on the real line R (or some subset of R) into R. The representation of such functions graphically as curves in the xy plane is a familiar and useful device.

Example 3.2

(i)   The correspondence

$$y = 1/(1 + x^2), \quad x \in R$$

defines a function from R into R. The graph is

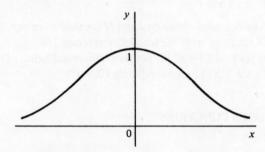

Figure 3.2(a)   Graph of $y = 1/(1 + x^2)$

(ii)   The correspondence

$$y = \ln x, \quad x > 0$$

defines a function on $(0, \infty) \subset R$ with range R.

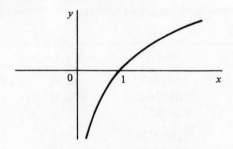

Figure 3.2(b)   Graph of $y = \ln x$

(iii)   Polynomials are expressions which, as we saw in chapter 2, are easy to compute and are widely used in numerical computation. We shall restrict consideration to polynomials with real coefficients; thus we take the expression

$$p_n(x) = a_0 + a_1 x + \ldots + a_n x^n$$

as defining a function from R into R. Polynomials are often used to approximate other functions which are more difficult to evaluate. The most commonly used polynomials are: constant, linear, quadratic, cubic ($n = 0, 1, 2, 3$). The graph of a polynomial of degree 4 has been plotted in figure 2.1. □

We see from figure 3.2(a) that, for a given function, there may be points $y \in R$ which are not values of the function for any $x \in R$. In fact, the function $f(x) = 1/(1 + x^2)$ has a maximum value of 1 at $x = 0$; there are no points x for which $f(x) > 1$ and in addition no points x for which $f(x) \leq 0$.

Definition 3.3

If a function f is defined on an interval I and there exists a real number M such that

$$f(x) \leq M \text{ for all } x \in I$$

then f is said to be *bounded above* on I and M is called an *upper bound*. (*Lower bound* is defined similarly with the inequality reversed.) □

The function $f(x) = 1/(1 + x^2)$ is bounded above and below. The function $f(x) = \ln x$ in figure 3.2(b) is unbounded on $(0, \infty)$.

3.2   CONTINUOUS FUNCTIONS

We now consider the function

$$f(x) = \frac{\ln x}{x - 1}$$

which is real-valued for all $x > 0$ except $x = 1$. At $x = 1$ we have $\ln x = 0$ and $x - 1 = 0$; hence $\ln x/(x - 1)$ is undefined. However, the function is defined at points arbitrarily close to $x = 1$. This is illustrated in the following graph and table, which were produced using the Pascal function ln and procedure graf from the mathlib library.

x	$\ln x$	$x - 1$	$f(x)$
0.50	−0.6931472	−0.50	1.386294
0.90	−0.1053605	−0.10	1.053605
0.99	−0.0100503	−0.01	1.005034
1.01	0.0099503	0.01	0.995033
1.10	0.0953102	0.10	0.953102
1.50	0.4054651	0.50	0.810930

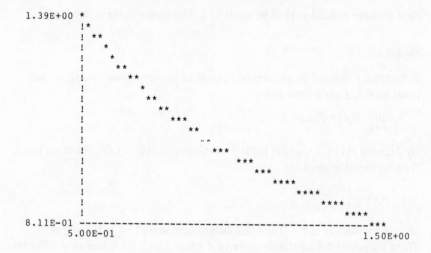

```
1.39E+00 *
         !*
         ! **
         !   *
         !     *
         !      **
         !       **
         !         *
         !          **
         !            **
         !              ***
         !                 **
         !                   ..^
         !                      ***
         !                         ***
         !                            ***
         !                               ****
         !                                  ****
         !                                     ****
         !                                        ****
8.11E-01 ----------------------------------------------------***
       5.00E-01                                                   1.50E+00
```

Figure 3.3   Graph of $f(x) = \ell n\, x/(x - 1)$ over $[0.5, 1.5]$

We have not attempted to evaluate $f(1)$ in our program; the result would be a runtime failure. From the table of values, however, it appears that $f(x)$ 'tends to' a definite value as x 'tends to' 1. We conjecture that the limiting value is 1. We say that $f(x)$ has a *limit* of 1 at $x = 1$, and we write

$$\lim_{x \to 1} f(x) = 1$$

It is important to understand that we cannot claim to have proved this by examining computer output†. The reader is advised to consult a textbook on analysis for further information (see reading list at the end of this chapter).

To code the function $f(x) = \ell n\, x/(x - 1)$ we would write

```
FUNCTION f ( x : real ) :  real ;
  BEGIN
  IF x = 1.0 THEN
    f := 1.0
  ELSE
    IF x > 0.0 THEN
      f := ln(x) / (x-1)
    ELSE  (* x <= 0.0 *)
      (* the function is undefined *)
      ..........
  END
```

†A function $f(x)$ is said to have the limit l at x_0 if, given any $\epsilon > 0$, we can find a quantity $\delta > 0$ such that $|f(x) - l| < \epsilon$ for all x in $(x_0 - \delta, x_0 + \delta)$, with the possible exception of $x = x_0$. From the table we can check particular cases, such as $|f(x) - 1| < 0.01$ for x satisfying $0.99 < x < 1.01$ (except $x = 1$).

Here we have defined $f(1)$ to be equal to 1. The reason is the following.

Definition 3.4

A function f defined on an interval I is said to be *continuous on I* if, at each point $x_0 \in I$, f has a limit and

$$\lim_{x \to x_0} f(x) = f(x_0) \quad \square$$

By defining $f(1) = 1$ we have made f continuous on $(0, \infty)$. On the other hand, the step function given by

$$g(x) = \begin{cases} -1, x \leq 0 \\ 1, x > 0 \end{cases}$$

has a discontinuity at $x = 0$ and this cannot be removed by redefining $g(0)$. There are points $\pm \delta$ arbitrarily close to $x = 0$ at which the values of g differ by 2; the function does not possess a limit at $x = 0$, hence it cannot be made continuous.

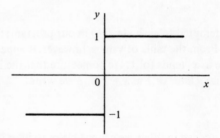

Figure 3.4   Graph of the step function $g(x)$

The Bisection Method

A basic problem of numerical computation is, for a given function f, to solve the equation $f(x) = 0$, that is, to find one or more *roots* α such that $f(\alpha) = 0$. If the function is known to be continuous on a certain interval, we can make use of the following theorem to search for real roots.

Theorem 3.1   (Intermediate value theorem)

If a real-valued function f is continuous on a closed interval $[a, b]$, then f takes on every value between $f(a)$ and $f(b)$ in (a, b).

[A consequence of this is that if $f(a)$ and $f(b)$ have opposite signs, that is, $f(a) f(b) < 0$, then there is at least one point $\alpha \in (a, b)$ at which $f(\alpha) = 0$.] $\square$

If we can find a suitable interval $[a, b]$ such that $f(a) f(b) < 0$, and if f is continuous, then we know that f has a root α in (a, b). We can narrow the interval known to contain α by calculating $\overline{x} = (a + b)/2$, then updating a or b as follows

IF $f(a) f(\overline{x}) < 0$ THEN
    $b := \overline{x}$
ELSE
    $a := \overline{x}$

Thus the width of the interval is halved. The reduced interval again qualifies under the intermediate value theorem, and the process can be repeated, apparently indefinitely. This is known as the *bisection method*.

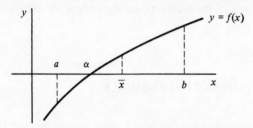

Figure 3.5   One step of the bisection method

Example 3.3

For $c > 0$ and n a positive integer, the equation

$$x^n - c = 0$$

has a positive real root

$$\alpha = c^{1/n}$$

If we can solve the equation we shall have computed the nth root of c. The expression $x^n - c$ is a polynomial, and it is known that polynomials are continuous on R. It follows that we can use the bisection method. If $0 < c < 1$, a suitable starting interval is $[0, 1]$; if $c > 1$, we can solve the equation $x^n - 1/c = 0$, which has a root in $(0, 1)$, and take the reciprocal of the result. The output from a program which solves the problem for $c = 0.2$ and $n = 5$ is now given.

```
5th root of  0.2000  using bisection method

        0.0000        1.0000
>>      0.5000        1.0000
        0.5000        0.7500  <<
>>      0.6250        0.7500
>>      0.6875        0.7500
>>      0.7187        0.7500
        0.7137        0.7344  <<
        0.7187        0.7266  <<
>>      0.7227        0.7266
>>      0.7246        0.7266
        0.7246        0.7256  <<

root is  0.7251  correct to 3 figures
```

At each step one of the bounds moves, as indicated by the symbols $\gg$, $\ll$. The root, $0.2^{1/5} = 0.724779\ldots$, is bracketed by the bounds at every step. The program is designed to stop when three correct figures are obtained because the bisection method, although very 'robust', is very slow! (Four more steps would be required to obtain the next correct figure.) □

The bisection method illustrates in a simple way the importance of mathematical analysis to numerical computation. The intermediate value theorem guarantees that the bisection method will work for any continuous function provided that a suitable starting interval can be found. We shall see in chapters 5 and 6 that there are better (faster and more general) methods for solving equations, although some of the most modern algorithms are still based in part on the bisection idea.

3.3   DERIVATIVES AND POLYNOMIALS

Rules for differentiating polynomials and other expressions will be familiar from calculus, as will techniques for differentiating products, quotients and 'functions of a function'. All such formulae depend ultimately on the concept of limit.

Definition 3.5

A function f defined on an open interval I is said to be *differentiable* at a point $x_0 \in I$ if the limit

$$\lim_{h \to 0} \frac{f(x_0 + h) - f(x_0)}{h}$$

exists. This limit is called the *derivative* of f at x_0. □

We recall that the derivative of f at x_0, if it exists, defines the slope of the tangent to the curve $y = f(x)$, as indicated in the following diagram.

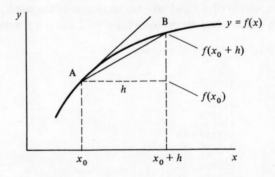

Figure 3.6   The tangent at A and the chord AB of the curve $y = f(x)$

If the derivative of f exists at all points in a subinterval $J \subseteq I$, then the values of the derivative define a function from J into R. This function is denoted by $f' : J \rightarrow R$. Since f' is a function, f' itself may have a derivative (second derivative of f), and so on. We use the notation

$$f', f'', \ldots, f^{(n)}, \ldots$$

for first, second and higher derivatives. If $f^{(n)}$ exists in a certain interval, then f is said to be *n times differentiable* in that interval (in context we shall often say that f is 'sufficiently differentiable').

It will sometimes be necessary to consider functions of several variables. For a function $f(x, y)$ of two variables, the *partial derivative with respect to x* at (x_0, y_0) is defined as

$$\lim_{h \to 0} \frac{f(x_0 + h, y_0) - f(x_0, y_0)}{h}$$

The partial derivative, if it exists, is interpreted as the rate of change of f as x varies with y held fixed. The corresponding derivative function (a function of x and y) is denoted by $\partial f / \partial x$ or f_x. [The *partial derivative with respect to y* is defined in similar fashion.]

Derivatives of Polynomials

We recall from chapter 2 that Horner's scheme is an efficient method of evaluating a polynomial

$$p_n(x) = \sum_{i=0}^{n} a_i x^i$$

at a specified point $x = \bar{x}$ (this method is coded as function polsum in section 2.2). We now show how to extend Horner's scheme to calculate the derivative at $\bar{x}, p_n'(\bar{x})$. This will be useful in algorithms for solving polynomial equations, to be discussed in chapter 5.

If $n = 0$ it is clear that $p_n(\bar{x}) = a_0$ and $p_n'(\bar{x}) = 0$. If $n > 0$, we shall show that $p_n(x)$ can be written in the form

$$p_n(x) = (x - \bar{x}) q_{n-1}(x) + r \tag{3.1}$$

where

$$q_{n-1}(x) = \sum_{i=0}^{n-1} b_i x^i \quad \text{(a polynomial of degree } n - 1)$$

and r is a constant remainder. Since, setting $x = \bar{x}$ in equation 3.1, we find $p_n(\bar{x}) = r$, our aim in the first instance is to determine r.

Equating coefficients of x^n, x^{k+1} and x^0 on both sides of equation 3.1, we obtain respectively

$$a_n \quad = b_{n-1}$$
$$a_{k+1} = b_k - \bar{x}\, b_{k+1}, \ 0 \leqslant k \leqslant n-2$$
$$a_0 \quad = -\bar{x}\, b_0 + r$$

These equations allow the coefficients of q_{n-1} to be calculated in the following order (this establishes the validity of equation 3.1)

$$\left.\begin{array}{l} b_{n-1} = a_n \\ b_k \quad = \bar{x}\, b_{k+1} + a_{k+1}, \ \ k = n-2, \ldots, 0 \\ r \quad\;\; = \bar{x}\, b_0 + a_0 \end{array}\right\} \tag{3.2}$$

The value of the polynomial, $p_n(\bar{x})$, is then given by r. This is precisely Horner's scheme as stated in section 2.1.

Now differentiating equation 3.1 we have

$$p_n'(x) = q_{n-1}(x) + (x - \bar{x})\, q_{n-1}'(x)$$

and setting $x = \bar{x}$

$$p_n'(\bar{x}) = q_{n-1}(\bar{x})$$

This shows that to determine $p_n'(\bar{x})$ (our original objective), we can instead determine $q_{n-1}(\bar{x})$. The coefficients of q_{n-1} have been calculated as the b_k values in equation 3.2. All that is required is to apply Horner's scheme again to $b_{n-1}, \ldots, b_0$.

Example 3.4

The extended Horner's scheme is coded in a Pascal procedure as follows.

```
PROCEDURE polval (      n : index ;
                    VAR a : coeff ;
                    x : real   ;
                    VAR val, grad : real ) ;
(* This procedure evaluates the polynomial
      p(x) = a[0] + a[1]*x + ... + a[n]*x^n
   and its derivative p'(x) for a specified value of x and
   n >= 0 using extended Horner's scheme. The value of p(x)
   is returned in val; the value of p'(x) is returned in grad *)

   VAR
      b, c : real ;
      i    : index ;

   BEGIN
   b := a[n] ;  c := 0.0 ;
   (* b is used to evaluate the polynomial,
      c is used to accumulate the derivative *)
   FOR i := n-1 DOWNTO 0 DO
      BEGIN
      c := c*x + b ;
      b := b*x + a[i]
      END ;
   val := b ;  grad := c
   END (* polval *)
```

Procedure polval should be compared with function polsum. The only addition is a new variable c which is used to accumulate the value of $q_{n-1}(\bar{x})$. □

3.4 LINEAR APPROXIMATIONS

We now quote a theorem on differentiable functions which is very useful in analysing numerical algorithms.

Theorem 3.2 (Mean value theorem for derivatives)

If a function f is continuous on a closed interval $[x_0, \overline{x}]$ and differentiable in $(x_0, \overline{x})$, then there is at least one point $\xi \in (x_0, \overline{x})$ at which

$$f'(\xi) = \frac{f(\overline{x}) - f(x_0)}{\overline{x} - x_0} \quad \square$$

The theorem has a simple geometrical interpretation. If a straight line is drawn through two points A, B on the curve $y = f(x)$, then there is some point ξ between A and B at which the tangent to the curve is parallel to the chord AB.

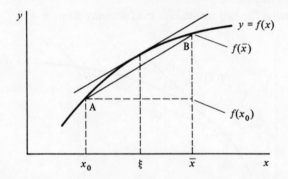

Figure 3.7   Geometrical interpretation of the mean value theorem

The mean value theorem may be written as

$$f(\overline{x}) = f(x_0) + (\overline{x} - x_0)\, f'(\xi), \quad \xi \in (x_0, \overline{x}) \tag{3.3}$$

This tells us the value of the function f at $\overline{x}$ in terms of the value at x_0 and the derivative at some point ξ, $x_0 < \xi < \overline{x}$. The value $f(x_0)$ is often known in applications, but the precise location of ξ is not known. In order to use equation 3.3 to 'predict' $f(\overline{x})$, we must be able to approximate $f'(\xi)$. This may be done in two ways.

(i)   *One point*
If the derivative of f is given at a certain point, usually x_0, then we take

$$f'(\xi) \approx f'(x_0)$$

Substituting into equation 3.3 we obtain

$$f(\overline{x}) \approx f(x_0) + (\overline{x} - x_0)\, f'(x_0)$$

(ii)   *Two points*

If the value of f is given at x_0 and at a second point x_1, then we can use this information instead of the derivative. Given $f(x_0) = f_0, f(x_1) = f_1$, we take

$$f'(\xi) \approx \frac{f_1 - f_0}{x_1 - x_0}$$

Then from equation 3.3,

$$f(\bar{x}) \approx f_0 + \left(\frac{\bar{x} - x_0}{x_1 - x_0}\right)(f_1 - f_0)$$

Rearranging, we obtain the symmetrical form

$$f(\bar{x}) \approx \left(\frac{\bar{x} - x_1}{x_0 - x_1}\right) f_0 + \left(\frac{\bar{x} - x_0}{x_1 - x_0}\right) f_1 \qquad (3.4)$$

This clearly gives the correct answer at $\bar{x} = x_0$ and $\bar{x} = x_1$, as the reader can check by substitution.

We summarise the two methods in the following diagram.

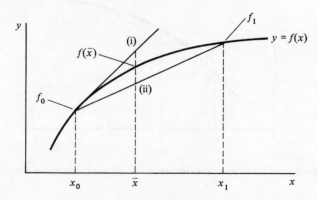

Figure 3.8   Linear approximations to $f(x)$

Method (i) amounts to following along the tangent line from (x_0, f_0) until the tangent intersects the ordinate $x = \bar{x}$. Method (ii) is equivalent to fitting a straight line (linear polynomial) through two points on the curve and evaluating this polynomial at $\bar{x}$. The approximation in equation 3.4 expresses an important idea known as *linear interpolation*.

Example 3.5

Calculate two approximations to $\sin 18°$.

Method (i)

$$\bar{x} = \pi/10; \quad \text{take } x_0 = 0$$
$$f(x) = \sin x, \quad f(x_0) = 0$$
$$f'(x) = \cos x, \quad f'(x_0) = 1$$

Then

$$f(\pi/10) \approx 0 + (\pi/10 - 0) \times 1$$
$$= 0.3142$$

Method (ii)

$$\bar{x} = \pi/10; \quad \text{take } x_0 = 0, \ x_1 = \pi/6$$
$$f_0 = \sin 0 = 0$$
$$f_1 = \sin (\pi/6) = 0.5$$

Then

$$f(\pi/10) \approx \left(\frac{\pi/10 - \pi/6}{0 - \pi/6} \right) 0 + \left(\frac{\pi/10 - 0}{\pi/6 - 0} \right) 0.5$$

$$= \quad 0.3000$$

($\sin 18° = 0.3090$ correct to 4 decimal places.) □

3.5 INTERPOLATION

In method (ii) of the previous section we have seen how to fit a straight line (linear polynomial) through two points $(x_0, f_0), (x_1, f_1)$ on a curve $y = f(x)$. The straight line is used in place of the function f to approximate $f(\bar{x})$ at an arbitrary point $\bar{x}$ (usually $\bar{x} \in (x_0, x_1)$). The curve $y = f(x)$ is, of course, in general non-linear, and so there is an *error*. We might expect to obtain a better approximation by using a polynomial of higher degree; for example, a quadratic polynomial has three coefficients, and these can be chosen so that the polynomial passes through three points $(x_i, f_i), i = 0, 1, 2$, on the curve.

To see how the method can be extended in this way we consider the expression in equation 3.4 for linear interpolation,

$$p_1(\bar{x}) = \left(\frac{\bar{x} - x_1}{x_0 - x_1} \right) f_0 + \left(\frac{\bar{x} - x_0}{x_1 - x_0} \right) f_1$$

We have written $p_1(\bar{x})$ to emphasise that this is a polynomial of degree 1 in x evaluated at a specified point $x = \bar{x}$. The expression consists of two terms with the following properties:

when $\bar{x} = x_0$, $p_1(x_0) = 1 \times f_0 + 0 \times f_1$
when $\bar{x} = x_1$, $p_1(x_1) = 0 \times f_0 + 1 \times f_1$

Taken together, these ensure that

$$p_1(x_0) = f_0, \quad p_1(x_1) = f_1$$

as required.

To fit a quadratic $p_2(x)$ through three points (x_i, f_i), $i = 0, 1, 2$, we try by analogy three terms

$$p_2(x) = \ell_0(x) f_0 + \ell_1(x) f_1 + \ell_2(x) f_2$$

$$= \frac{(x - x_1)(x - x_2)}{(x_0 - x_1)(x_0 - x_2)} f_0 + \frac{(x - x_0)(x - x_2)}{(x_1 - x_0)(x_1 - x_2)} f_1$$

$$+ \frac{(x - x_0)(x - x_1)}{(x_2 - x_0)(x_2 - x_1)} f_2 \qquad (3.5)$$

Each term is of degree 2 in x, and it is easy to check by substitution that

$$p_2(x_0) = f_0, \quad p_2(x_1) = f_1, \quad p_2(x_2) = f_2$$

as required for the interpolating quadratic. We then take, as an approximation to $f(\bar{x})$,

$$f(\bar{x}) \approx p_2(\bar{x})$$

Example 3.6

Calculate an approximation to $\sin 18°$ by quadratic interpolation based on the known values of $\sin x$ at $x = 0, 30°$ and $45°$.

$$\bar{x} = \pi/10$$
$$x_0 = 0, \quad x_1 = \pi/6, \quad x_2 = \pi/4$$
$$f_0 = 0, \quad f_1 = 0.5, \quad f_2 = 1/\sqrt{2}$$

We evaluate $p_2(\bar{x})$ from formula 3.5. Since $f_0 = 0$, the first term is 0; the second term is

$$\frac{(\pi/10 - 0)(\pi/10 - \pi/4)}{(\pi/6 - 0)(\pi/6 - \pi/4)} \times 0.5 = 0.54$$

It is convenient to set out the calculations systematically as follows

i	x_i	$\bar{x} - x_i$	$\ell_i(\bar{x})$	f_i	$\ell_i f_i$
0	0	$\pi/10$	0.24	0	0
1	$\pi/6$	$-\pi/15$	1.08	0.5000	0.5400
2	$\pi/4$	$-3\pi/20$	-0.32	0.7071	-0.2263

$$\Sigma = 0.3137$$

Thus

$$f(\pi/10) \approx p_2(\pi/10)$$

$$= \sum_{i=0}^{2} \ell_i f_i = 0.3137$$

This should be compared with example 3.5, where two different linear approximations were calculated. □

In principle we need not stop at quadratic interpolation. It can be shown that for a set of points (x_i, f_i), $i = 0, 1, \ldots, n$, where $n \geqslant 0$ and the x_i values are all distinct, there exists a unique polynomial $p_n(x)$ of degree $\leqslant n$ which passes through all $n + 1$ points. This is called *the interpolating polynomial of degree n*. One way of expressing this polynomial is as a generalisation of equation 3.5

$$p_n(x) = \sum_{i=0}^{n} \ell_i(x) f_i$$

where

$$\ell_i(x) = \prod_{\substack{j=0 \\ j \neq i}}^{n} \frac{x - x_j}{x_i - x_j} \quad \text{for } 0 \leqslant i \leqslant n$$

(3.6)

Formula 3.6 is *Lagrange's formula* for the interpolating polynomial. The notation indicates that each $\ell_i(x)$ is a product of n factors of the form stated (the factor corresponding to $j = i$, $(x - x_i)/(x_i - x_i)$, is omitted). Again it is quite easy to check that

$$p_n(x_i) = f_i, \quad 0 \leqslant i \leqslant n$$

which shows that the polynomial passes through all $n + 1$ points as required.

The interpolating polynomial can be used as in example 3.6 to calculate approximate values of a function based on a suitable set of data (usually up to five or six known function values). In practice, the values of a function are sometimes available only at a finite set of discrete points. An example is a table of experimental results recording the dependence of some physical quantity on a single independent variable. Only a finite number of experimental measurements can be made, and the experimenter then wishes to use the results to test a theoretical model or to make predictions. Interpolation has a limited practical application for this purpose. However, Lagrange's formula is not very suitable for automatic computation, and more convenient and flexible formulations are used in practice.

For our purposes the chief importance of the interpolating polynomial is theoretical. The fact that it exists and can be expressed by Lagrange's formula allows us to derive numerical approximation rules and error formulae for the estimation of definite integrals (see chapter 8).

The *error* of the interpolating polynomial at $x = \bar{x}$ is defined as

$$f(\bar{x}) - p_n(\bar{x})$$

that is, the exact function value minus the estimate. The error is obviously non-zero in general, and it is important both theoretically and practically to estimate its size. We quote the following theorem (for a proof see, for example, Phillips and Taylor (1973), chapter 4).

Theorem 3.3

If $p_n(x)$ is the polynomial defined by equation 3.6, and if the function f has the properties

(i) f is continuous and has continuous derivatives up to $f^{(n)}$ on an interval $[a, b]$ containing all the points $x_0, x_1, \ldots, x_n$ and $\overline{x}$

(ii) $f^{(n+1)}$ exists at all points in (a, b)

then there is a point $\xi \in (a, b)$ such that

$$f(\overline{x}) - p_n(\overline{x}) = \frac{(\overline{x} - x_0) \ldots (\overline{x} - x_n)}{(n + 1)!} f^{(n+1)}(\xi) \quad \square$$

The precise location of the point ξ, which depends on $\overline{x}$, is not known. Often the best we can hope to do is to determine the largest possible value of $|f^{(n+1)}(\xi)|$ and hence obtain an *error bound*. In some cases we can go further and calculate more refined upper and lower bounds on the error. In example 3.6 we have $f(x) = \sin x$ and $n = 2$, hence

$$f^{(n+1)}(\xi) = -\cos \xi$$

The quantities $\overline{x} - x_i$ for $\overline{x} = \pi/10$ are tabulated in example 3.6; the error of $p_2(\pi/10)$ can be expressed by theorem 3.3 as

$$E = \frac{(\pi/10) \times (-\pi/15) \times (-3\pi/20)}{3!} (-\cos \xi)$$

$$= -0.0052 \cos \xi$$

The point ξ lies in $(0, \pi/4)$, thus

$$0.7071 < \cos \xi < 1.0$$

By substitution in the error expression we find that

$$-0.0052 < E < -0.0037$$

In example 3.6 we have determined $p_2(\pi/10) = 0.3137$, so adding on the error E we can finally claim

$$0.3085 < \sin \pi/10 < 0.3100$$

3.6 SEQUENCES AND SERIES

The bisection method was used in example 3.3 to compute the nth root of a positive number c. The results consist of successive pairs of real numbers a, b which bracket the root with increased accuracy at each step. We now formalise this by saying that the bisection method generates two *sequences* of real numbers, denoted by $\{a_k\}, \{b_k\}, k = 0, 1, \ldots$. The sequences are such that the width $b_k - a_k$ can be made as small as desired for k sufficiently large, and since

$c^{1/n} \in (a_k, b_k)$ for all k, both sequences approach arbitrarily closely to $c^{1/n}$.

This is an idea of wide applicability in numerical computation. If we wish to compute a quantity α which cannot be determined directly, we approach it indirectly by setting up one or preferably two sequences, designed so that by computing a sufficient number of terms we can approximate α to any desired accuracy. The concept is worth a precise definition.

Definition 3.6

A sequence of real numbers $\{u_k\}$, $k = 0, 1, \ldots$, is said to *converge to the limit l* (symbolically, $\lim_{k \to \infty} u_k = l$) if, given any $\epsilon > 0$, we can find an integer N such that

$$|u_k - l| < \epsilon$$

for all $k > N$. $\square$

In less precise language, for k sufficiently large, u_k approaches arbitrarily closely to l. This summarises the behaviour of the sequences $\{a_k\}$, $\{b_k\}$ in the bisection method,

$$\lim_{k \to \infty} a_k = \lim_{k \to \infty} b_k = c^{1/n}$$

The width of the interval in the bisection method is halved at each step, $b_k - a_k = (b_0 - a_0)/2^k$, and it is easy to see that the underlying sequence $\{1/2^k\}$, $k = 0, 1, \ldots$, converges to 0. Given any $\epsilon > 0$, we can choose N so that $1/2^N \leqslant \epsilon$; then for any $k > N$

$$|1/2^k - 0| < \epsilon \quad \text{(compare definition 3.6)}$$

To what extent such limiting processes can be realised on finite computing machines in finite time is a question we must consider carefully when we design and code our algorithms.

In different applications we may be concerned with sequences not only of real numbers but also of complex numbers, vectors, matrices or functions. Such sequences may also converge to a limit; definition 3.6 then has to be generalised. We shall deal with some of these cases as we meet them in later chapters.

An *infinite series* is defined in terms of a sequence as an expression of the form

$$a_0 + a_1 + a_2 + \ldots$$

where $\{a_k\}$, $k = 0, 1, \ldots$, is any sequence of numbers. The finite sum

$$s_k = a_0 + a_1 + \ldots + a_k, \quad k \geqslant 0$$

is called the *kth partial sum* of the series. It is natural to ask whether an infinite series has a 'sum to infinity'.

Definition 3.7

An infinite series $a_0 + a_1 + a_2 + \ldots$ with kth partial sum s_k is said to be
convergent if the sequence of partial sums $\{s_k\}$, $k = 0, 1, \ldots$, converges to a
limit. The limit $\lim\limits_{k \to \infty} s_k = s$, if it exists, is called the *sum of the series.* □

Example 3.7

Prove that the infinite series

$$1 + \frac{1}{1!} + \frac{1}{2!} + \ldots$$

is convergent.

The partial sums up to s_6 are shown in the following table.

k	0	1	2	3	4	5	6	...
s_k	1.0000	2.0000	2.5000	2.6667	2.7083	2.7167	2.7181	...

(to 4 decimal places)

To prove that the series is convergent, we must show that the sequence $\{s_k\}$,
$k = 0, 1, \ldots$, converges to a limit. Since the numbers s_k increase monotonically,
we need only prove that the sequence is bounded above (see Courant and John
(1965), chapter 1). For any $k > 1$

$$s_k = 1 + 1 + \frac{1}{2} + \frac{1}{2 \times 3} + \ldots + \frac{1}{2 \times 3 \times \ldots \times k}$$

$$\leqslant 2 + \frac{1}{2} + \frac{1}{2^2} + \ldots + \frac{1}{2^{k-1}}$$

$$= 2 + \frac{1/2 - 1/2^k}{1 - 1/2} < 3$$

Since $s_k < 3$ for all k, the sequence is bounded; $\lim\limits_{k \to \infty} s_k$ therefore exists, and
the series is convergent.

(The sum is $e = 2.71828\ldots$.) □

In practice we can only compute a partial sum of an infinite series; we truncate
the series at a term a_k and take the partial sum s_k as an approximation to the
sum s. The remainder, $s - s_k$, is in this context known as the *truncation error*. In
order to make use of s_k, and also to decide when to stop, we must be able to
assess the size of the truncation error. For some commonly occurring series this
is easy to do (see Dahlquist and Björck (1974), chapter 3). For example, an
alternating series of the form

$$1 - 1/2 + 1/3 - \ldots$$

where the terms alternate in sign and decrease in absolute value to zero, has a remainder which is easily bounded at each step. In this case it can be proved that

$$|s - s_k| < |a_{k+1}|$$

that is, the absolute value of the truncation error is less than the absolute value of the first neglected term.

3.7 TAYLOR'S SERIES

In section 3.5 we described a method of approximating a function f by the interpolating polynomial, a polynomial of degree n which matches the value of f at $n + 1$ points, $x_0, x_1, \ldots, x_n$. A different approach is to consider a single point x_0 and to construct a polynomial which matches the function f and its derivatives $f', \ldots, f^{(n)}$ all at x_0. For this to be possible f must be sufficiently differentiable at x_0. We use the abbreviation $f_0^{(j)}$ to denote the derivatives $f^{(j)}(x_0)$, $1 \leqslant j \leqslant n$.

The polynomial

$$\pi_n(x) = f_0 + (x - x_0)f_0' + \frac{(x - x_0)^2}{2!} f_0'' + \ldots + \frac{(x - x_0)^n}{n!} f_0^{(n)} \quad (3.7)$$

has the necessary properties. $\pi_n(x)$ is a polynomial of degree n in x (the quantities $x_0, f_0, f_0', \ldots, f_0^{(n)}$ are all constants). By substitution we see that $\pi_n(x_0) = f_0$; by differentiation and substitution, $\pi_n'(x_0) = f_0'$; in general, $\pi_n^{(j)}(x_0) = f_0^{(j)}, 1 \leqslant j \leqslant n$. Thus, provided the necessary derivatives are available, we can form a polynomial which matches the function f up to the nth derivative at x_0.

Example 3.8

Construct the polynomial $\pi_n(x)$ for the function $f(x) = e^{-x}$ about $x_0 = 0$.
We have $f^{(j)}(x) = (-1)^j e^{-x}$. At $x_0 = 0$,

$$f_0 = 1$$
$$f_0' = -1$$
$$\cdot$$
$$\cdot$$
$$f_0^{(n)} = (-1)^n$$

The polynomial 3.7 becomes

$$\pi_n(x) = 1 - x + \frac{x^2}{2!} - \ldots + (-1)^n \frac{x^n}{n!} \quad \square$$

We now consider a point

$$\bar{x} = x_0 + h$$

with $h \neq 0$ and, as in the case of the interpolating polynomial, we take $\pi_n(\bar{x})$ as an approximation to $f(\bar{x})$. (In what follows we simplify the description by assuming $h > 0$, although h may equally well be negative.

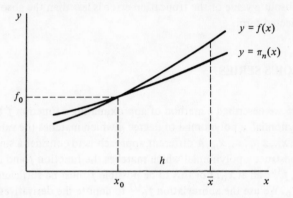

Figure 3.9   Polynomial approximation $\pi_n(x)$ to $f(x)$ about x_0

The *error* of the approximation at $\bar{x}$ is defined as

$$R_n(h) = f(\bar{x}) - \pi_n(\bar{x})$$
$$= f(x_0 + h) - \pi_n(x_0 + h)$$

The error is clearly non-zero in general (it is zero when $h = 0$, and we would expect it to be smallest for h small; also the error depends on the behaviour of f between x_0 and $\bar{x}$). In practical applications we wish to estimate the size of the error, and for theoretical purposes a formula for $R_n(h)$ is essential. We quote the following theorem, which is proved in textbooks of classical analysis.

Theorem 3.4   (Taylor's theorem)

If $\pi_n(x)$ is the polynomial defined by equation 3.7, and if the function f has the properties

   (i) f is continuous and has continuous derivatives up to $f^{(n)}$ on an interval $[x_0, \bar{x}]$
   (ii) $f^{(n+1)}$ exists at all points in $(x_0, \bar{x})$
   then there is a point $\xi \in (x_0, \bar{x})$ such that

$$f(\bar{x}) - \pi_n(\bar{x}) = \frac{(\bar{x} - x_0)^{n+1}}{(n+1)!} f^{(n+1)}(\xi) \quad \square$$

By substituting $\pi_n(\bar{x})$ from equation 3.7, with $h = \bar{x} - x_0$, we can write Taylor's theorem in the form of a very useful equation

$$f(x_0 + h) = f_0 + hf_0' + \frac{h^2}{2!} f_0'' + \ldots + \frac{h^n}{n!} f_0^{(n)} + \frac{h^{n+1}}{(n+1)!} f^{(n+1)}(\xi) \tag{3.8}$$

where

$$x_0 < \xi < x_0 + h$$

Formula 3.8 is the *truncated Taylor's series with remainder* (the case $n = 0$ gives theorem 3.2, the mean value theorem).

Since ξ in equation 3.8 is an unknown point which depends on h, we cannot calculate the error (or remainder) $R_n(h)$ exactly. What we can attempt to do is to determine the largest possible value of $|f^{(n+1)}(\xi)|$ and hence obtain an *error bound*.

Example 3.9

Obtain an error bound for the approximation $\pi_n(x)$ to e^{-x} in example 3.8. We have

$$f(x) = e^{-x}$$
$$f^{(n+1)}(\xi) = (-1)^{n+1} e^{-\xi}$$

and since $x_0 = 0, h = x$. By Taylor's theorem the error is

$$R_n(x) = \frac{x^{n+1}}{(n+1)!} (-1)^{n+1} e^{-\xi}$$

If we restrict the approximation to $x > 0$, then $\xi \in (0, x)$ and it follows that $e^{-\xi} < e^0 = 1$ (note we cannot assume that $\xi = x$). The error bound is then

$$|R_n(x)| < \frac{x^{n+1}}{(n+1)!}$$

For example, if we wish to approximate e^{-1}, we can evaluate the cubic $\pi_3(x)$ at $x = 1$,

$$\pi_3(1) = 1 - 1 + \frac{1}{2!} - \frac{1}{3!}$$

(This is a truncated form of an alternating series as referred to at the end of section 3.6.) We find that $\pi_3(1) = 0.3333$, and the error bound is $1/4! = 0.0417$. ($e^{-1} = 0.3679$ correct to 4 decimal places.) □

The reader will have met Taylor's series about $x_0 = 0$ (also known as Maclaurin's series) for $e^{\pm x}$, $\sin x$, $\cos x$ and certain other functions. All such series, together with the error when the series is truncated after a finite number of terms, are derived from equation 3.8.

Example 3.10

(i)   The truncated Taylor's series for $\sin x$ about $x_0 = 0$ is

$$x - \frac{x^3}{3!} + \frac{x^5}{5!} - \ldots + (-1)^{n-1} \frac{x^{2n-1}}{(2n-1)!}$$

The truncation error is

$$R_{2n}(x) = (-1)^n \frac{x^{2n+1}}{(2_n+1)!} \cos \xi, \quad \xi \in (0,x)$$

and since $\max_{\xi \in R} |\cos \xi| = 1$ we have

$$|R_{2n}(x)| \leq \frac{|x^{2n+1}|}{(2n+1)!}$$

The series can be used to calculate $\sin x$ for any x. It appears that for large x many terms may have to be taken to make the error acceptably small (but see the exercises for this chapter).

(ii)   The truncated Taylor's series for $\ln x$ about $x_0 = 1$ is

$$(x-1) - \frac{(x-1)^2}{2} + \frac{(x-1)^3}{3} - \ldots + (-1)^{n-1} \frac{(x-1)^n}{n}$$

The truncation error is

$$(-1)^n \frac{(x-1)^{n+1}}{(n+1)\xi^{n+1}}$$

where $x \in (0, \infty)$ and ξ lies between 1 and x. This series can be used to prove that the function $\ln x/(x-1)$, discussed in section 3.2, has a limit of 1 at $x = 1$ (see exercises). □

Order of Approximation

The remainder term in the truncated Taylor's series of equation 3.8 can be considered in a more general way. If the derivative $f^{(n+1)}$ is bounded on $(x_0, x_0 + h)$, that is, if there exists a real number M such that $|f^{(n+1)}(\xi)| \leq M$ for all $\xi \in (x_0, x_0 + h)$, then the remainder in equation 3.8 is bounded,

$$|R_n(h)| \leq Mh^{n+1}/(n+1)!$$

We express this by writing

$$R_n(h) = 0\,(h^{n+1})$$

and we say that the Taylor's series truncated at the term in h^n is an *approximation of order* h^{n+1} to $f(x_0 + h)$. In general, the order of approximation shows the dependence of the error on h as $h \to 0$. A similar terminology will be used for other approximation formulae which depend on a non-zero quantity h and in which the error can be made arbitrarily small as $h \to 0$ (examples are formulae for derivatives and definite integrals).

An important application of Taylor's series is to obtain formulae for derivatives, which are used in the numerical solution of differential equations. One approximation to $f'(x_0)$, suggested by definition 3.5 for the derivative, is given by the *difference quotient*

$$\frac{f(x_0 + h) - f(x_0)}{h}, \quad h \neq 0$$

This approximation is $0(h)$ provided that f'' exists and is bounded. It is easy to obtain an approximation for the derivative which is $0(h^2)$ (see exercises). The higher the order of approximation, the more rapidly the error decreases as h is reduced. For example, if we have two formulae for $f'(x_0)$, one $0(h)$ and the other $0(h^2)$, then halving the stepsize h will approximately halve the error in the first case but will reduce the error by a factor of approximately 4 in the second case. In general we seek to use high-order approximation formulae because they can give high accuracy for small computational effort. The drawback is that the function must be 'sufficiently differentiable', that is, possess derivatives up to a high order in a suitable interval.

Example 3.11

Use Taylor's series to establish the approximations

$$f'(x_0) = \frac{f(x_0 + h) - f(x_0)}{h} + 0(h)$$

$$f''(x_0) = \frac{f(x_0 + h) - 2f(x_0) + f(x_0 - h)}{h^2} + 0(h^2)$$

(i) $\quad f(x_0 + h) = f_0 + hf_0' + \frac{h^2}{2} f''(\xi), \ \xi \in (x_0, x_0 + h)$

$\Rightarrow \dfrac{f(x_0 + h) - f(x_0)}{h} = f'(x_0) + 0(h)$

(ii) $\quad f(x_0 + h) = f_0 + hf_0' + \frac{h^2}{2} f_0'' + \frac{h^3}{6} f_0^{(3)} + \frac{h^4}{24} f^{(4)}(\xi_1), \ \xi_1 \in (x_0, x_0 + h$

$\quad f(x_0 - h) = f_0 - hf_0' + \frac{h^2}{2} f_0'' - \frac{h^3}{6} f_0^{(3)} + \frac{h^4}{24} f^{(4)}(\xi_2), \ \xi_2 \in (x_0 - h, x_0)$

$\Rightarrow \dfrac{f(x_0 + h) - 2f(x_0) + f(x_0 - h)}{h^2} = f''(x_0) + 0(h^2)$

(We assume that f is sufficiently differentiable, that is, the derivatives required in the error terms exist and are bounded.) □

Further Reading

There are many introductory textbooks on mathematical analysis, some of which are more readable than others. We mention here Burkill (1970), Courant and John (1965) and Apostol (1957).

Among the many modern texts on general numerical analysis may be recommended Phillips and Taylor (1973), Conte and de Boor (1980) (with FORTRAN programs), and the stimulating book by Dahlquist and Björck (1974).

EXERCISES

3.1   Explain why
(i) $\pi \in (3.14, 3.15)$ but $\pi \notin [3.142, 3.143]$;
(ii) $[0.0, 1.0] \not\subset R_m$, where R_m is the set of machine numbers of a digital computer;
(iii) the function $f: (-\infty, c) \to R$ given by $f(x) = \ln(1 - x)$ for $x < 1$, $f(x) = 0$ for $x \geqslant 1$, is bounded below if and only if $c < 1$;
(iv) if f is defined as $f: R \to [0, \infty)$ with $f(x) = x^2$, then the inverse relation of f from $[0, \infty)$ to R is not a function.

3.2   Following exercise 1.5, develop a program to locate a value of x which satisfies the equation $x^n - c = 0$, where $c > 0$ and n is a positive integer (you may use the mathlib function poweri). The program should determine x correct to 4 significant figures (see also example 3.3). Test the program by calculating $0.125^{1/3}$ and $32.0^{1/5}$, then use it to compute the fifth and tenth roots of 0.2. Compare the results with those of function powerr.

3.3   Derive the computational equations 3.2 for Horner's scheme explicitly from equation 3.1 in the case when $p_n(x)$ is a cubic polynomial. For the polynomial $p(x) = x^3 - 3x - 10$, determine $p(3)$ and $p'(3)$ by 'hand running' procedure polval. Explain what happens in polval if N = 0 or N = 1.

3.4   Calculate three approximations to $\cos 36°$ using
(i)   the mean value theorem (based on $\cos 60°$);
(ii) linear interpolation (based on $\cos 0, \cos 60°$);
(iii) quadratic interpolation (based on the values of $\cos x$ at $x = 0$, $60°, 90°$).
Explain, with a diagram, why the three results are different. Use the result of part (iii) and theorem 3.3 to obtain bounds on $\cos 36°$. How does this compare with the correct value?

3.5   Given that $p(x)$ and $q(x)$ are two polynomials, both of degree $\leqslant n$, which pass through the points (x_i, y_i), $i = 0, 1, \ldots, n$ (where $n \geqslant 0$ and the x_i

values are all distinct), prove that $p(x)$ and $q(x)$ must be the same polynomial. (You may assume that a polynomial of degree n has at most n real zeros, that is, points at which its value is zero. Then prove that the difference $p(x) - q(x)$ is identically zero.) Deduce that the interpolating polynomial given by equations 3.6 is unique. What is the unique polynomial of degree $\leqslant 3$ which passes through the points $(1, 2), (2, 3), (3, 4), (4, 5)$?

3.6   Use Taylor's series to calculate $e^{1.1}$ correct to 4 decimal places, given $e = 2.71828$ to 5 decimal places. Choose a suitable base point for the expansion and explain how you decide the number of terms to take.

3.7   Construct the truncated Taylor's series with remainder about $x_0 = 0$ for $\ln(1 - x), x < 1$. Calculate two approximations to $\ln 1.2$ by taking the first three terms and the first four terms of the series, and obtain bounds on the error. Compare the error bounds with the actual error.

3.8   Use Taylor's series to calculate $\sin 666$ (radians) correct to 5 decimal places. The series for $\sin x$ is given in example 3.10(i). If you substitute $x = 666$ directly, you will need a large number of terms and encounter severe problems of overflow and cancellation. Devise a way round this using the periodicity of the sine function.

3.9   Establish the Taylor's series for $\ln x$ given in example 3.10(ii). Use the series to prove that $\ln x/(x - 1)$ has a limit of 1 at $x = 1$, paying particular attention to the behaviour of the remainder term as $x \to 1$.

3.10   Use Taylor's series expansion about x_0 to verify the approximations to the first derivative of a function $f(x)$

$$f'(x_0) = \frac{f(x_0 + h) - f(x_0 - h)}{2h} + 0\,(h^2)$$

$$f'(x_0) = \frac{-3f(x_0) + 4f(x_0 + h) - f(x_0 + 2h)}{2h} + 0\,(h^2)$$

Obtain expressions for the truncation error, assuming that f is sufficiently differentiable.

4 Real Numbers, Machine Numbers and Errors

> If you please, ma'am, it was a very little one.
>
> Captain Marryat, *Midshipman Easy*

Calculations with real numbers, whether performed by computer or by hand, are subject to *numerical error*. This form of error is unavoidable in numerical computation and should not be confused with 'gross errors' arising from programming mistakes or from incorrect design or misuse of algorithms. On a computer, numerical error is fundamentally *machine dependent:* its size depends on the way in which real numbers are represented in the computer. This is an important consideration for us to bear in mind when developing a mathematical library.

In this chapter we identify the sources of numerical error and assess the effects on our computations. An understanding of this is essential if we are to design algorithms which are reliable (capable of producing results of predictable accuracy) and adaptable to a variety of computers. For algorithms which are to be coded as library routines, this is at least as important as reducing the computational cost.

4.1 ERRORS AND THEIR SOURCES

Perhaps the most obvious source of error in numerical computation is *data error*. Experimental data involving real numbers (measurements of length, time etc.) are always subject to uncertainty. The irreducible component of experimental error is usually considered to be random, and results are often stated in the form $x = \bar{x} \pm \epsilon$, indicating that the true value x lies in an interval $(\bar{x} - \epsilon, \bar{x} + \epsilon)$ with given probability. Errors with a similar effect are inherent even in mathematical constants such as π and e. Although such constants are defined precisely in mathematical terms, their representation requires an infinite sequence of digits

$$\pi = 3.1415926\ldots, \quad e = 2.7182818 \ldots \text{ (in base 10)}$$

To use these in a calculation we must approximate the values to a finite number of digits, for example,

$\pi \approx 3.142$ (rounded to 3 decimal places/4 significant figures)
$e \approx 2.71828$ (rounded to 5 decimal places/6 significant figures)

The consequence is a data error.

More generally, there are almost always idealisations made in setting up a mathematical model before any computation can begin. The discrepancy between the model and the physical system is a source of error which can be regarded as error in the data of the problem. (Mathematical models and their relation to the real world are investigated in applied mathematics, physics, engineering, etc.)

Of more immediate concern to us in numerical computation are the fundamental limitations imposed by the finite nature of the computing medium at our disposal. Every possible means of computation is subject to the restrictions of finite time and finite space. Finite time gives rise to *truncation error* and finite space to *rounding error*.

Truncation Error

Many numerical methods express the desired result as the limit of an infinite sequence or the sum of an infinite series. In practice only a finite number of terms can be computed; the consequence of neglecting the remaining terms is an error known as *truncation error* (also called in some contexts 'discretisation error'). The underlying theory often provides us with an estimate of this error or a bound on its size. Most of the numerical methods discussed in this book have corresponding formulae for truncation error. Examples are

(i)   The truncated Taylor's series with remainder for the value of a function or a derivative (section 3.7).
(ii)   Numerical approximations to a definite integral in terms of a number of strips fitted under a curve. The number of strips must be finite, and in general there is an error (example 1.1 and chapter 8).
(iii)   The interval $[a_k, b_k]$ in the bisection method, which brackets the root of an equation to any desired accuracy. The width of the interval is never exactly zero, and the result is a margin of error in the computed root (sections 3.2, 3.6 and chapter 6).

Where appropriate, our algorithms are designed to allow the user to control the truncation error by specifying a tolerance within which the error of the result will normally lie.

Rounding Error

Computers usually express numbers internally to base 2 or a related base. As a binary fraction the number 1/10, for example, is an infinite recurring sequence

$$1/10 = 0 \times \frac{1}{2} + 0 \times \frac{1}{2^2} + 0 \times \frac{1}{2^3} + 1 \times \frac{1}{2^4} + 1 \times \frac{1}{2^5} + \ldots$$

$$= (.0\ 0011\ 0011\ \ldots)_2$$

It follows that $1/10$ cannot be represented exactly in a computer with a binary or related base. Most real numbers cannot be represented exactly in any base; they must necessarily be approximated to fit the space available for storing digits (in a computer this is governed by the *word length*). The approximation is usually performed by *rounding*, for example, $1/10 \approx (0.0001100110)_2$ rounded to 10 binary places. (This is quite a poor approximation to $1/10$, equivalent to $(0.0996\ldots)_{10}$; large computers typically employ 25 to 50 binary digits.)

The effect of machine rounding will be discussed in more detail later. It is important at this stage to understand that *rounding error* can arise not only in the original data but at every step of a calculation. For example, the product of two t-digit numbers consists of $2t$ or $2t - 1$ digits; normally this must be rounded to t digits before the next step of the calculation can proceed. In long computations the accumulation of rounding error can have a serious effect on the final answer, and even in short calculations there are pitfalls to be avoided. The consequences for computer arithmetic will be discussed in section 4.4.

4.2   MEASUREMENT AND CONTROL OF ERROR

In numerical computation we frequently wish to carry out an iterative process (calculation of successive approximations) until the result is correct to a specified number of decimal places. In order to be able to terminate the iteration when the desired accuracy is attained, we require a quantitative measure of numerical error and a precise statement of what is meant by 'number of correct decimal places'.

Absolute Error

Definition 4.1

If $\bar{x}$ is an approximation to a real number x, then the *absolute error* of $\bar{x}$ is $|\bar{x} - x|$. □

Given that the approximation $\bar{\pi} = 3.142$ is obtained by rounding π to 3 decimal places, we can infer that the true value π satisfies

$$3.1415 \leqslant \pi < 3.1425\dagger$$

†There is an ambiguity in rounding if the first neglected digit is 5. We follow the convention of most computer systems which employ rounding and always round *up* in such cases.

that is

$$\overline{\pi} - 0.5 \times 10^{-3} \leqslant \pi < \overline{\pi} + 0.5 \times 10^{-3}$$

Generalising, we see that if $\overline{x}$ is obtained by rounding a positive number x to t decimal places, then

$$\overline{x} - 0.5 \times 10^{-t} \leqslant x < \overline{x} + 0.5 \times 10^{-t}$$

Rounding may introduce an error of up to half a unit in the last decimal place. It follows that the absolute error in rounding a number to t decimal places is bounded according to

$$|\overline{x} - x| \leqslant 0.5 \times 10^{-t} \tag{4.1}$$

Conversely, we may use equation 4.1 to define what is meant by 't correct decimal places'. An approximation $\overline{x}$ to x is said to be *correct to t decimal places* if t is the largest integer such that the inequality 4.1 is satisfied. We refer to this as an *absolute error test*. For $\pi = 3.14159\ldots$ and $\overline{\pi} = 3.142$, the absolute error is $|\overline{\pi} - \pi| = 0.41 \times 10^{-3} < 0.5 \times 10^{-3}$, and we see that $\overline{\pi}$ is indeed correct to 3 decimal places.

Many library routines in numerical computation include a tolerance parameter which the user must set to control the accuracy (truncation error) of the output. In the mathlib library this parameter will often be an absolute error tolerance denoted by tol. The computation is continued until a bound or estimate of the truncation error is, if possible, less than tol in absolute value. If results are required correct to t decimal places, a suitable value for tol can be calculated from tol $= 0.5 \times 10^{-t}$.

Example 4.1

In example 3.10 we saw that $\sin x$ can be approximated by

$$x - \frac{x^3}{3!} + \ldots + (-1)^{n+1} \frac{x^{2n-1}}{(2n-1)!}$$

with a truncation error bounded in absolute value by the next term in the series

$$\frac{|x|^{2n+1}}{(2n+1)!}$$

Write a Pascal function to compute $\sin x$ for $|x| \leqslant \pi/2$ to an absolute accuracy specified by tol.

```
FUNCTION sine ( x, tol : real ) :  real ;
(* Computes the sine of x to absolute accuracy tol *)

    VAR
      xsq, term, sum          : real ;
      i (* summation index *) : integer ;
```

```
BEGIN
xsq := sqr(x) ;
term := x ;    (* initialise the variables *)
sum := 0.0 ;   (* used in the summation     *)
i := 1 ;
REPEAT
    sum := sum + term ;     (* add term in x^i to partial sum *)
    i := i + 2 ;            (* update the index *)
    term := -term * xsq/((i-1)*i)  (* compute next term *)
UNTIL abs(term) < tol ;  (* absolute error test *)
sine := sum  (* truncation error is bounded in absolute value
                by tol *)
END (* sine *)
```

The function can be used to compute $\sin x$ correct to t decimal places if a value $\text{tol} = 0.5 \times 10^{-t}$ is calculated in the main program. A summary of results for $x = \pi/2$ is given ($\sin \pi/2 = 1.0000...$).

t	$\text{tol} = 0.5 \times 10^{-t}$	sine (x, tol)	$|\text{error}| <$ tol
2	$0.5E-2$	1.004525	$0.4525E-2$
3	$0.5E-3$	0.999843	$0.157E-3$
4	$0.5E-4$	1.000004	$0.04E-4$

Notes

(i)   The user may set an accuracy which is unattainable on the computer (an extreme case would be $\text{tol} = 0.0$). An important problem is to detect this and return the best possible answer for the computer in use (see section 4.3).

(ii)   The error is bounded at each step as the calculation proceeds; it would not be practicable to determine the number of steps in advance from the condition $|x|^{2n+1}/(2n+1)! < \text{tol}$.

(iii)   The function is designed to compute $\sin x$ only for $|x| \leqslant \pi/2$, although the series expansion is valid for arbitrary x. A simple modification is required to make the algorithm work reasonably efficiently for any x (see exercises)†. □

Relative Error

Definition 4.2

If $\bar{x}$ is an approximation to a non-zero real number x, then the *relative error* of $\bar{x}$ is $|\bar{x} - x|/|x|$. □

It is convenient in calculations with very large or very small numbers, such as $c = 485165195.4...(= e^{20})$, to work in terms of *significant figures* rather than decimal places. We may then write $\bar{c} = 4852 \times 10^5$ (rounded to 4 significant

†The standard functions sin, cos, ln etc. in programming languages such as Pascal are evaluated not in terms of Taylor polynomials but in terms of other approximating functions which distribute the truncation error more evenly over a base interval (see Dahlquist and Björck (1974), chapter 4).

figures). To measure the closeness of $\overline{c}$ to c, the appropriate quantity is usually not the absolute error but the error relative to c. The relative error is

$$\frac{|\overline{c} - c|}{|c|} = 0.72 \times 10^{-4} \ (0.0072 \text{ per cent})$$

We now ask how we can test if an approximation $\overline{x}$ to x is *correct to t significant figures*. By this we shall mean that if x is expressed in normalised floating-point form

$$x = f \times 10^m \text{ where } 0.1 \leqslant |f| < 1.0$$

and similarly

$$\overline{x} = \overline{f} \times 10^m$$

then $\overline{f}$ should agree with f correct to t decimal places. For example, with c and $\overline{c}$ above

$$c = 0.4851651954 \times 10^9$$
$$\overline{c} = 0.4852 \times 10^9$$

hence $|\overline{f} - f| = 0.35 \times 10^{-4}$, and we conclude that $\overline{c}$ agrees with c correct to 4 significant figures.

A convenient test for accuracy to t significant figures is the *relative error test*

$$\frac{|\overline{x} - x|}{|x|} \leqslant 0.5 \times 10^{-t} \tag{4.2}$$

From equation 4.2 it follows that

$$\frac{|\overline{f} - f|}{|f|} \leqslant 0.5 \times 10^{-t}$$

hence

$$|\overline{f} - f| < 0.5 \times 10^{-t} \text{ (since } |f| < 1)$$

This shows that $\overline{f}$ agrees with f to at least t decimal places, and therefore $\overline{x}$ has at least t correct significant figures.

If the user of a library routine requires results correct to t significant figures, then a relative error test should be used. If, as in some of the mathlib routines, only an absolute error parameter tol is provided, this will necessitate an order of magnitude $\overline{x}$ to be known for the result. A suitable value for tol can then be calculated from

$$\text{tol} = 0.5 \times 10^{-t} \times |\overline{x}|$$

and this can be supplied as an absolute error tolerance to the library routine.

4.3   MACHINE NUMBERS AND MATHLIB CONSTANTS

To accommodate as wide a range of real numbers as possible, digital computers use the *normalised floating-point representation*. A non-zero real number x is expressed in floating-point form as

$$x = f \times \beta^m, \quad 1/\beta \leqslant |f| < 1.0$$

where f is the *fractional part* (in base β), β is the *base* (usually 2, sometimes 10 or 16), and m is the *exponent*. The quantities f and m may be positive or negative (actual computers have special arrangements for representing negative numbers and other details which need not concern us). The number 0.0, which cannot be 'normalised', is treated as a special case. For all other numbers the condition on $|f|$ ensures that the first digit of the fractional part is non-zero, and the representation is then unique. Thus, 1/10 would be expressed on a 10 binary digit floating-point computer as

$$(0.1100110011)_2 \times 2^{-3} \quad \text{(see section 4.1)}$$

Specific parts of the computer word (or words) are allocated to hold the fraction f and the exponent m of a floating-point number. Since the word length of any computer is finite, this imposes two fundamental restrictions on real numbers.

(i)   *Finite range*   The exponent m is restricted to a range $-L \leqslant m \leqslant M$, where L and M are machine constants (L is typically 64 or 128 on binary machines, and $M = L - 1$). Thus there is a maximum real number representable on the computer and, less obviously, a minimum positive real number. Any attempt to store numbers lying outside this range results in *floating-point overflow* or *underflow*.

(ii)   *Finite precision*   The fraction f has a length of exactly t digits, where t is a machine constant (t is typically 25 to 50 binary digits on large computers). Thus any number x having more than t significant digits in base β cannot be represented exactly, even if x lies within the permissible range of real numbers for the computer.

Together these two restrictions imply that only a finite subset of real numbers has an exact representation on the computer. This subset, $R_m \subset R$, is called the set of *machine numbers*.

Example 4.2

In a model computer with decimal base ($\beta = 10$), a floating-point number is represented as a signed 4-digit fraction ($t = 4$) and a signed 1-digit exponent. What are the maximum and minimum positive real numbers representable on the computer, and what would be a suitable value for the mathlib constant smallreal?

The machine numbers are of the form

$$\pm 0.d_1 d_2 d_3 d_4 \times 10^{\pm m}$$

where $d_1 \neq 0$ and $0 \leqslant m \leqslant 9$. The fractional part ranges from 0.1000 to 0.9999 and the exponent part from 10^{-9} to 10^9. The maximum real number is therefore 0.9999×10^9. The minimum positive real number is 0.1000×10^{-9}. The library constant smallreal, defined in chapter 2 as 100 times the smallest positive real number, is

smallreal = 1.0E-8 □

We now give a Pascal program to estimate experimentally the minimum positive real number on any computer. In this program the variable x will ultimately underflow, and the result on most systems will be a runtime failure. The last number written before this occurs is the smallest positive real number to within a factor of 10. From this the constant smallreal required in the mathlib library can be calculated (a more accurate value can be obtained, if desired, from machine documentation).

```
PROGRAM minr ( output ) ;

VAR
   x : real ;

BEGIN
writeln ('minimum positive real number') ;
x := 1.0 ;
REPEAT
   writeln (x:10) ;
   x := x/10
UNTIL x = 0.0    (* last number written is the minimum
                    positive real number to within
                    a factor of 10 *)

END .
```

Output on the Amdahl V/7 (abbreviated)

```
minimum positive real number
 1.00E+00
 1.00E-01
 ........
 ........
 ........
 1.00E-73
 1.00E-74
 1.00E-75
```

Since the set R_m of machine numbers is finite, whereas the set R of real numbers is infinite, how can we perform computations with real numbers? A real number x (provided it lies within the permissible range for the computer) is represented by a floating-point machine number $f\ell(x)$. Naturally we would like to minimise the error $|f\ell(x) - x|$, and this can be achieved if $f\ell(x)$ is chosen to agree with x correct to t significant digits. This means that $f \ell (x)$ is obtained by *rounding* x to the nearest machine number†.

Example 4.3

In the 4-decimal digit computer of example 4.2, what are the machine number representations of the real numbers 3.14159, −0.012345, 45678.9, 45684.5?

x	$f \ell (x)$	$\lvert \bar{f} - f \rvert$	$\lvert f \ell (x) - x \rvert / \lvert x \rvert$
3.14159	0.3142×10^1	0.41×10^{-4}	0.13×10^{-3}
−0.012345	-0.1235×10^{-1}	0.50×10^{-4}	0.41×10^{-3}
45678.9	0.4568×10^5	0.11×10^{-4}	0.02×10^{-3}
45684.5	0.4568×10^5	0.45×10^{-4}	0.10×10^{-3}

The absolute error in the fractional part of $f \ell (x)$ is listed as $\lvert \bar{f} - f \rvert$, and we note that $\lvert \bar{f} - f \rvert \leqslant 0.5 \times 10^{-4}$ ($f \ell (x)$ agrees with x correct to 4 significant decimal digits). We also note that the *relative error*, $\lvert f \ell (x) - x \rvert / \lvert x \rvert$, is less than 0.5×10^{-3} in all cases. □

The maximum relative error incurred in representing real numbers by machine numbers is an important constant in numerical computation called the *relative precision*. We derive a formula for the relative precision as follows.

A non-zero real number x can be expressed in normalised floating-point form in base β as $x = f \times \beta^m$. To represent x on a machine with base β and a precision of t digits, the fractional part f is rounded to t digits; the result is the machine number $f \ell (x) = \bar{f} \times \beta^m$. (If f happens to round up to 1.0, then the machine number is actually $(1/\beta) \times \beta^{m+1}$; this can be written as $1.0 \times \beta^m$, and the derivation is unaffected.) By a generalisation of equation 4.1 we have

$$\lvert \bar{f} - f \rvert \leqslant \tfrac{1}{2} \beta^{-t}$$

and since f is normalised, $\lvert f \rvert \geqslant 1/\beta$. The relative error is therefore

$$\frac{\lvert f \ell (x) - x \rvert}{\lvert x \rvert} = \frac{\lvert \bar{f} - f \rvert}{\lvert f \rvert} \leqslant \frac{\tfrac{1}{2} \beta^{-t}}{1/\beta} = \tfrac{1}{2} \beta^{1-t} \tag{4.3}$$

Thus the relative precision (maximum possible relative error in machine rounding) is given by $\tfrac{1}{2} \beta^{1-t}$.

Example 4.4

Determine the relative precision of the 4-decimal digit computer in example 4.2. Also determine the smallest positive number ϵ such that $f \ell (1 + \epsilon) \neq 1$. What value should be used for the mathlib constant rprec4 on this computer?

† An alternative occasionally used is *chopping*, that is, discarding all digits after the first t. This leads to faster implementations at the expense of larger errors in arithmetic operations.

Since $\beta = 10$ and $t = 4$, the relative precision is, from equation 4.3, 0.5×10^{-3} (this should be compared with the relative error in example 4.3). The smallest positive number ϵ such that $f\ell(1 + \epsilon) \neq 1$ is also 0.5×10^{-3}, since $f\ell(1.0 + 0.5 \times 10^{-3}) = 0.1001 \times 10^{1} \neq 1.0$. The library constant rprec4, defined in chapter 2 as 4 times the relative precision, is

    rprec4 = 2.0E − 3 □

We now give a Pascal program to estimate experimentally the relative precision of the computer. We use the observation in example 4.4 and determine the smallest positive number eps such that $1.0 + \text{eps} \neq 1.0$. The last number written by the program is the relative precision to within a factor of 2 (from this the library constant rprec4 can be calculated).

```
PROGRAM rprec ( output ) ;

VAR
    eps : real ;

BEGIN
writeln ('relative precision') ;
eps := 1.0 ;
REPEAT
    writeln (eps:10) ;
    eps := eps/2
UNTIL 1.0 + eps = 1.0    (* last number written is the relative
                            precision to within a factor of 2 *)
END .
```

Output on the Amdahl V/7 using REAL*8

```
relative precision
 1.00E+00
 5.00E-01
 ........
 ........
 ........
 8.88E-16
 4.44E-16
 2.22E-16
```

(this should be compared with the output from program minr).

In terms of equation 4.3 we can express the relation between a real number x and its machine approximation $f\ell(x)$ in a useful form. We have

$$\frac{f\ell(x) - x}{x} = \epsilon$$

where ϵ may be positive or negative and $|\epsilon| \leqslant \frac{1}{2}\beta^{1-t}$. Hence

$$f\ell(x) = x(1 + \epsilon), \quad |\epsilon| \leqslant u = \frac{1}{2}\beta^{1-t} \tag{4.4}$$

where u is the relative precision of the computer.

Minimum Tolerance in Error Tests

The iteration loop in the sine function (example 4.1) has the form

```
REPEAT
    sum := sum + term ;
    . . . . . . . . . .
UNTIL abs(term) < tol
```

where tol is a tolerance parameter set by the user. How can we guard against the user setting an impossible tolerance, for example, tol = 0.0? What accuracy is it sensible to ask for?

The above code is typical of many iterative processes which we meet in practice. A quantity x is computed by a formula

$$x_{new} := x_{old} + \delta$$

where δ is a correction calculated at each iteration. The truncation error, $|x - x_{new}|$, is known to be bounded by $|\delta|$, and the process is terminated when $|\delta| < $ tol.

In practice we must also take account of machine rounding in the computation of x_{new} from x_{old}. When the iterative process is executed on a computer, x_{new} is actually assigned the value of the machine number $f\ell(x_{old} + \delta)$, that is, $x_{new} := f\ell(x_{old} + \delta)$. From equation 4.4 we then have

$$x_{new} = (x_{old} + \delta)(1 + \epsilon), \quad |\epsilon| \leqslant u$$

where u is the relative precision. The possible size of the rounding error is therefore $|x_{old} + \delta| u \approx |x_{new}| u$. It is not meaningful to test for a truncation error smaller than this quantity; we must be sure to use a tolerance greater than this to control the iteration loop. We take, as a safe minimum value for the tolerance,

$$\text{tolmin} = 4 |x_{new}| u$$

$$= |x_{new}| \text{ rprec4}$$

Example 4.5

On the 4 decimal digit computer, what is the value of tolmin for (i) $x = 2.718$, (ii) $x = -0.1234 \times 10^{-7}$?

From example 4.4 we have rprec4 = 2.0×10^{-3}, hence

(i)   tolmin = $2.718 \times 2.0 \times 10^{-3}$
         = 5.436×10^{-3}

(ii)   tolmin = $0.1234 \times 10^{-7} \times 2.0 \times 10^{-3}$
         = 0.2468×10^{-10}

In case (ii) we meet a difficulty. The minimum positive real number representable on the computer is 1.0×10^{-10} (see example 4.2), and thus the value computed for tolmin would underflow. To avoid this we set tolmin to the value of the library constant smallreal

$$\text{tolmin} = 1.0 \times 10^{-8}$$

In fact, whenever $|x|$ rprec4 < smallreal, we define tolmin to be equal to smallreal. □

The value of tolmin depends on the current iterate, and therefore tolmin has to be recomputed at each iteration. We provide a utility function in the mathlib library to compute the minimum tolerance for any argument.

```
FUNCTION tolmin ( x : real ) :   real ;
(* Computes the minimum tolerance to be used in error tests.
   Library constants used:   smallreal, rprec4 *)

   BEGIN
   IF abs(x) > smallreal/rprec4 THEN
      tolmin := abs(x)*rprec4
   ELSE
      tolmin := smallreal   (* abs(x)*rprec4 may underflow *)
   END (* tolmin *)
```

We can now modify the stopping criterion in the sine function as follows

```
REPEAT
   sum := sum + term ;
   ..........
UNTIL abs(term) < tol + tolmin(sum)
```

To compute $\sin x$ to full machine accuracy we would simply set tol = 0.0.

4.4   ARITHMETIC ON A 4-DIGIT COMPUTER

As we have seen in section 4.3, real numbers are represented on a computer by machine numbers. The elementary arithmetic operations $+, -, *, /$, which are the 'building blocks' for all calculations with real numbers, can be performed only on machine numbers, and the result of each operation must normally be stored in the computer as a machine number. We might expect this to have some effect on the accuracy of our calculations. We now examine the effects in some important cases.

We consider a 4-decimal digit floating-point computer. As in section 4.3 this is a model computer which illustrates the essential features of floating-point arithmetic†. The computer is provided with a double-length (8-digit) *accumulator* in which the arithmetic operations $+, -, *, /$ are performed. Thus, for example, the product of 0.1234×10^2 and 0.4567×10^{-1} appears in the accumulator as 0.05635678×10^1.

When one operation has been performed and before the accumulator is free for the next operation, the current result must be placed in the machine's store rounded to 4 significant figures. If the exact result of an operation on machine numbers a and b is denoted by $a \circ b$, then from equation 4.4 the result actually stored is the machine number

$$fl\,(a \circ b) = (a \circ b)\,(1 + \epsilon), \quad |\epsilon| \leqslant 0.5 \times 10^{-3} \tag{4.5}$$

At every such stage a rounding error may be introduced.

To simulate 4-digit arithmetic on an actual computer, we use a library

† A similar discussion will be found in Cohen (1973), chapter 1.

function which rounds the value of a real variable or expression to 4 significant decimal digits.

```
FUNCTION roundt ( x : real ) :  real ;
Returns x rounded to 4 significant decimal digits.
```

(The code of roundt is given in appendix C.)

Summation of Series

The 4-digit computer performs addition of two floating-point numbers according to the following steps.

(i)   The number which is smaller in absolute value is loaded into the accumulator.

(ii)   The exponents of the two numbers must be made equal: the exponent of the smaller number is increased if necessary, and to compensate for this the fractional part is shifted right in the accumulator.

(iii) The larger number is added into the accumulator.

(iv)   The result is stored as a machine number rounded to 4 decimal digits.

One unexpected result of this process (which is fairly typical of computer addition) is that the sum of a set of numbers may depend on the order in which the numbers are added.

Example 4.6

Compute $a + b + c$ on the 4-digit computer, given $a = 0.6472 \times 10^0$, $b = 0.4685 \times 10^{-4}$, $c = 0.3297 \times 10^{-4}$.

We shall add the numbers twice, once in the order given and a second time in reverse order (following the rules stated above).

(i)	STORE	ACCUMULATOR
b	0.4685×10^{-4} $\longrightarrow$	$0.0000\ 4685 \times 10^0$
$a + b$		$0.6472\ 4685 \times 10^0$
$fl\,(a + b)$	0.6472×10^0 $\longleftarrow$	
c	0.3297×10^{-4} $\longrightarrow$	$0.0000\ 3297 \times 10^0$
$fl\,(a + b) + c$		$0.6472\ 3297 \times 10^0$
$fl\,(fl\,(a + b) + c)$	0.6472×10^0 $\longleftarrow$	

The result is equal to a (b and c, both much smaller than a, have been lost).

(ii) STORE ACCUMULATOR

c 0.3297×10^{-4} ⟶ $0.3297\,0000 \times 10^{-4}$

$b + c$ $0.7982\,0000 \times 10^{-4}$

$fl\,(b + c)$ 0.7982×10^{-4}

                                       $0.0000\,7982 \times 10^{0}$

$a + fl\,(b + c)$ $0.6472\,7982 \times 10^{0}$

$fl\,(a + fl\,(b + c))$ 0.6473×10^{0}

The result is *not* equal to a (b and c have combined together and the combined effect has not been lost). Backward summation in part (ii) is more accurate than forward summation in part (i). □

To show the same effect in a more realistic application, we consider the summation of a series containing a few comparatively large terms and many much smaller terms.

Example 4.7

Compute the sum of the series

$$1 + \frac{1}{2^2} + \frac{1}{3^2} + \ldots + \frac{1}{n^2} + \ldots$$

on the 4-digit computer. (The sum can be determined analytically; it is $\pi^2/6 = 1.6449\ldots$.)

We give a program to compute the partial sum s_n by forward and backward summation for $n = 20, 40, \ldots, 200$. A 4-digit computer is simulated by the library function roundt.

```
PROGRAM simsum ( input, output ) ;
(* This program computes the sum of the n terms (1/k)^2, k=1,..,n,
   using two different orders of summation:
   sumup (increasing order of k), sumdown (decreasing order of k).
   The computation is repeated for n = 20, 40,.., 200.  A 4-digit
   decimal computer is simulated by the library routine roundt *)

VAR
   i, k, n        : integer ;
   sumup, sumdown : real ;

FUNCTION roundt ( x : real ) :  real ;
   EXTERN ;

BEGIN
writeln ( '    n      forward sum     backward sum' ) ;
writeln ;
FOR i := 1 TO 10 DO
   BEGIN
   n := 20*i ;
   (* Compute the sum of n terms by forward and backward
      summation on the simulated 4-digit machine *)
```

```
   sumup := 0.0 ;  sumdown := 0.0 ;
   FOR k := 1 TO n DO
      BEGIN
      sumup := roundt ( sumup + roundt(1/sqr(k)) ) ;
      sumdown := roundt ( sumdown + roundt(1/sqr(n+1-k)) )
      END ;
   writeln (n:6, sumup:12:3, sumdown:14:3)
   END
END .
```

The results are

n	forward sum	backward sum
20	1.594	1.596
40	1.619	1.620
60	1.623	1.628
80	1.623	1.632
100	1.623	1.635
120	1.623	1.637
140	1.623	1.638
160	1.623	1.639
180	1.623	1.639
200	1.623	1.640

The forward sum does not progress beyond 1 correct decimal place (this is because the small terms are added into the sum later and are lost as in example 4.6). The backward sum converges slowly to approximately 3 correct decimal places (the small terms combine together first and are not lost). □

To obtain the best accuracy in summing a large set of numbers which differ widely in magnitude, we should whenever possible accumulate the smallest numbers first†.

Cancellation Error

There is a possibility of serious error when two floating-point numbers are subtracted (or two numbers of opposite sign are added). If the numbers are nearly equal, the fractional part of the result will consist of a string of leading zeros followed by a few non-zero digits. When the result is normalised only the first few digits will be meaningful, and the answer may contain a large relative error which will be propagated in any subsequent calculation. This is known as *cancellation error*.

Example 4.8

Solve the quadratic equation $x^2 - 30x + 1 = 0$ on the 4-digit computer.

The roots are $15 \pm \sqrt{224}$. Assuming that $\sqrt{224}$ can be computed correct to 4 figures, we have

$$f\ell(15) \quad = 0.1500 \times 10^2$$
$$f\ell(\sqrt{224}) = 0.1497 \times 10^2$$

†However, this may not be a practical proposition. In the sine function (exampl 4.1) the number of terms is not known in advance, and so backward summation is not implemented.

Hence on the 4-digit computer

$$x_1 = 0.1500 \times 10^2 + 0.1497 \times 10^2 = 0.2997 \times 10^2$$
$$x_2 = 0.1500 \times 10^2 - 0.1497 \times 10^2 = 0.3000 \times 10^{-1}$$

The larger root, x_1, is correct to 4 figures, but x_2 has only one correct figure (it should be 0.03337...). The smaller root obtained on the 4-digit computer has a 10 per cent relative error caused by subtraction of nearly equal numbers. □

An algorithm which admits serious cancellation error should be reformulated. This can be done for the solution of quadratic equations by computing the numerically larger root x_1 in the usual way and determining the smaller root x_2 from the relation $x_2 = c/ax_1$. There is no subtraction and hence no cancellation error. In the example

$$x_2 = c/ax_1$$
$$= 0.1000 \times 10^1/0.2997 \times 10^2$$
$$= 0.3337 \times 10^{-1}$$

This idea can be incorporated in a quadratic equation solver (see example 1.4) as follows

```
t := sqr(b) - 4*a*c ;
IF t >= 0.0 THEN
   BEGIN  (* roots are real *)
   t := sqrt(t) ;
   (* Assign numerically larger root to x1 *)
   IF b > 0.0 THEN
      x1 := (-b - t) / (2*a)
   ELSE
      x1 := (-b + t) / (2*a) ;
   (* Compute the smaller root, avoiding zero division *)
   IF x1 <> 0.0 THEN
      x2 := c/(a*x1)
   ELSE
      x2 := 0.0
   END  (* real roots *)
ELSE
   (* roots are complex *)
   . . . . . . . . . .
```

4.5   ERROR PROPAGATION

So far we have considered the rounding error introduced when arithmetic operations are performed on machine numbers; the error depends on the precision of the computer and is governed by a relation of the form of equation 4.5. In general, however, the operands already contain data errors and rounding errors from previous calculations. We now ask how errors already present are *propagated* in further calculations.

To analyse error propagation we write

$$\bar{a} = a + \delta a, \quad \bar{b} = b + \delta b$$

where $\delta a, \delta b$ are the errors in the approximate values $\bar{a}, \bar{b}$. For addition or subtraction we have

$$\bar{a} \pm \bar{b} = (a \pm b) + (\delta a \pm \delta b)$$

If $\bar{S}$ is used to denote the computed result $\bar{a} \pm \bar{b}$ and S the correct result $a \pm b$, this may be written as

$$\bar{S} = S + (\delta a \pm \delta b)$$

Hence

$$|\bar{S} - S| = |\delta a \pm \delta b| \leqslant |\delta a| + |\delta b|$$

since each of the errors $\delta a, \delta b$ may be positive or negative. A bound on the *absolute error* of a sum or difference is given by the *sum of bounds* on the *absolute errors* of the operands.

For multiplication or division we write

$$\bar{a} = a(1 + r_a), \quad \bar{b} = b(1 + r_b)$$

where

$$r_a = \delta a/a, \quad r_b = \delta b/b$$

Then

$$\bar{a}\bar{b} = ab(1 + r_a)(1 + r_b)$$

and

$$\bar{a}/\bar{b} = (a/b)(1 + r_a)/(1 + r_b)$$

Using P and $\bar{P}$ to denote the correct and computed results respectively, we have in the case of multiplication

$$\bar{P} \approx P(1 + r_a + r_b)$$

hence

$$\frac{|\bar{P} - P|}{|P|} \approx |r_a + r_b| \leqslant \left|\frac{\delta a}{a}\right| + \left|\frac{\delta b}{b}\right|$$

For division

$$\bar{P} = P(1 + r_a)(1 + r_b)^{-1}$$
$$\approx P(1 + r_a)(1 - r_b) \approx P(1 + r_a - r_b)$$

hence

$$\frac{|\bar{P} - P|}{|P|} \approx |r_a - r_b| \leqslant \left|\frac{\delta a}{a}\right| + \left|\frac{\delta b}{b}\right|$$

A bound on the *relative error* of a product or quotient is given, to first-order approximation, by the *sum of bounds* on the *relative errors* of the operands.

These results are concerned with propagated error. We have ignored the entirely separate contribution of rounding error which would arise in actually performing the arithmetic operations on a computer, as discussed in section 4.4. This is justifiable provided

$$|\delta a/a|, |\delta b/b| \gg u$$

where u is the relative precision of the computer.

The bounds for the basic arithmetic operations are special cases of error propagated in a function. Error is, of course, propagated in general functions. If the value of $f(x)$ is required at $x = a$ but an approximate argument $x = \bar{a} \approx a$ is used instead, then

$$f(a) = f(\bar{a}) + (a - \bar{a}) f'(\xi)$$

where ξ is an unknown point lying between a and $\bar{a}$ (see the mean value theorem, section 3.4). In applications f is continuously differentiable, and we often take $f'(\xi) \approx f'(\bar{a})$. Thus the absolute error in the function value is

$$|f(\bar{a}) - f(a)| \approx |\delta a| \times |f'(\bar{a})|$$

where $\delta a = \bar{a} - a$. If a bound or estimate is known for $|\delta a|$, this can be used to estimate the error propagated in the function.

Example 4.9

A quantity a is measured in an experiment: $a = 0.98$ correct to 2 decimal places. The formula

$$f(x) = \frac{1}{\sqrt{(1 - x^2)}}$$

is to be evaluated at $x = a$, and this is calculated as $f(0.98) = 5.025$. What accuracy can be claimed for the answer?

In our notation we have $\bar{a} = 0.98$, $|\delta a| \leqslant 0.5 \times 10^{-2}$, and $|f(\bar{a}) - f(a)| \approx |\delta a| \times |f'(\bar{a})|$. Since $f(x) = (1 - x^2)^{-1/2}$, $f'(\bar{a}) = \bar{a}(1 - \bar{a}^2)^{-3/2} = 124.4$. Thus the possible error in $f(\bar{a})$ is estimated as

$$|f(\bar{a}) - f(a)| \lesssim 0.005 \times 125 = 0.625$$

that is

$$f(a) = 5.025 \pm 0.625$$

The error estimate is not unduly pessimistic, as the following table shows.

a	0.975	0.980	0.985	
$f(a)$	4.500	5.025	5.795	□

We see from example 4.9 that the error in the solution of a problem can sometimes be much greater than the error in the data we supply. If a small perturbation of the data can produce a large change in the solution, the problem is said to be *ill-conditioned*. Perturbations can arise effectively from rounding errors in the calculation as well as from data error. If an ill-conditioned problem cannot be reformulated, careful precautions must be taken to reduce rounding error in the calculation (see in particular chapter 7).

For a function $f(x, y, \ldots)$ of several variables the analogous result for error propagation is

$$|f(\bar{a}, \bar{b}, \ldots) - f(a, b, \ldots)|$$

$$\lesssim |\delta a| \times \left| \frac{\partial f}{\partial x} \right| + |\delta b| \times \left| \frac{\partial f}{\partial y} \right| + \ldots$$

where the partial derivatives are usually evaluated at the known point $(\bar{a}, \bar{b}, \ldots)$. This can be used to obtain an error estimate which, however, often greatly over-estimates the actual error; the formula assumes the worst case, with the individual errors reinforcing each other. More realistic estimates can be obtained by a statistical approach in which the errors $\delta a, \delta b, \ldots$ are treated as independent random variables (see, for example, Dahlquist and Björck (1974), chapter 2).

4.6   NUMERICAL INSTABILITY OF ALGORITHMS

Even if the problem to be solved is well-conditioned, a poorly designed algorithm can produce worthless results. The error in a sequence of calculations may grow to the extent of 'swamping' the true solution, so that the results computed are meaningless. An algorithm with this defect is said to be *numerically unstable*.

Example 4.10

Evaluate the integral

$$I_n = \int_0^1 x^n e^x \, dx, \quad n \geq 0$$

using a recurrence relation.

Integrating by parts we obtain

$$I_n = [x^n e^x]_0^1 - n \int_0^1 x^{n-1} e^x \, dx$$

$$= e - n I_{n-1} \text{ for } n \geq 1 \tag{4.6}$$

Since $I_0 = \int_0^1 e^x \, dx = e - 1$, we can develop a forward recurrence algorithm based on equation 4.6 to compute $\{I_1, I_2, \ldots\}$ up to any desired I_n. This is programmed in Pascal as follows

```
PROGRAM forrecur ( output ) ;

CONST
    e = 2.71828 ;    (* correct to 5 decimal places *)
    n = 12 ;         (* number of terms required *)
```

```
VAR
    ival : real ;
    k    : integer ;

BEGIN
writeln ('forward recurrence for integral of  x^n * exp(x)') ;
writeln ;
ival := e - 1.0 ;    (* initialise the sequence *)
FOR k := 1 TO n DO
    BEGIN
    ival := e - k*ival ;   (* apply recurrence relation *)
    writeln (k:8, '       ', ival:12:6)
    END
END .
```

The output is

```
forward recurrence for integral of  x^n * exp(x)
         1        1.000000
         2        0.718280
         3        0.563440
         4        0.464520
         5        0.395680
         6        0.344200
         7        0.308880
         8        0.247240
         9        0.493120
        10       -2.212920
        11       27.060400
        12     -322.006520     □
```

Are the results correct? It is easy to see that they are not. Since $x^n e^x > 0$ for $x > 0$, it follows that each I_n must be positive. Also

$$I_{n+1} - I_n = \int_0^1 (x^{n+1} - x^n)\, e^x \mathrm{d}x$$

and since $x^{n+1} - x^n < 0$ for $0 < x < 1$, we have $I_{n+1} - I_n < 0$, that is, $I_{n+1} < I_n$. Thus $\{I_0, I_1, I_2, \ldots\}$ should be a strictly *decreasing* sequence of positive numbers. The output has the correct general behaviour up to I_8; after this the numbers alternate in sign and increase rapidly in magnitude. This is typical of unstable behaviour. The forward recurrence algorithm is numerically unstable, and the results for $n > 8$ are worthless.

The recurrence relation is correct, so what is wrong? We note that the value used for e in the program is rounded to 5 decimal places; thus there is a data error in e and therefore in I_0. In fact,

$$\overline{I}_0 = I_0 + \epsilon_0, \quad |\epsilon_0| \leqslant 0.5 \times 10^{-5}$$

The following terms are then computed in succession

$$\overline{I}_1 = e - 1 \times \overline{I}_0 = (e - I_0) - \epsilon_0 = I_1 - \epsilon_0$$
$$\overline{I}_2 = e - 2 \times \overline{I}_1 = (e - 2I_1) + 2\epsilon_0 = I_2 + 2\epsilon_0$$

and in general

$$\overline{I}_k = e - k \times \overline{I}_{k-1} = I_k + (-1)^k k!\, \epsilon_0$$

(We ignore the further error introduced by the incorrect value of e at each step, since this complicates the details without altering the sense.)

The error propagated to I_n is

$$|\epsilon_n| \sim n! \times 0.5 \times 10^{-5}$$

For $n = 9$ this is already approximately 1; at this point the error begins to explode and swamp the true solution. For $n = 12$ the possible size of the error is

$$12! \times 0.5 \times 10^{-5} \approx 2000$$

This accounts for almost the entire 'value' calculated for I_{12}!

It is clear that numerical instability is a property of the algorithm, not of the problem (instability has nothing to do with the computer − similar results would be obtained by hand). It is little use taking a more accurate value for e, since the factorial increases so rapidly. However, a modified algorithm will produce correct results for this problem. The relation in equation 4.6 can be expressed as a backward recurrence

$$I_{n-1} = (e - I_n)/n \quad \text{for } n \geqslant 1$$

Since the sequence is known to be decreasing, we can take, for example, $I_{12} = 0$ and use this to compute $I_{11}, I_{10}, \ldots, I_0$. The backward recurrence algorithm is numerically stable, as can be shown by a similar analysis (note the division by n). The initial error in I_{12} and any further errors incurred are quickly damped away.

```
backward recurrence for integral of x^n * exp(x)

        11          0.226523
        10          0.226523
         9          0.249176
         8          0.274345
         7          0.305492
         6          0.344684
         5          0.395599
         4          0.464536
         3          0.563436
         2          0.718281
         1          0.999999
         0          1.718281
```

Although the starting value for I_{12} was arbitrarily taken as 0, after a few steps the terms are correct (I_0 agrees with e − 1 correct to 5 decimal places).

EXERCISES

4.1 Modify function sine (section 4.2) to compute $\sin x$ to an absolute accuracy tol for any real argument (see exercise 3.8). Use the library function tolmin discussed in section 4.3 to make the algorithm robust. Test the function on $\sin 666$ (radians) to the full accuracy of the computer, and compare with the answer given by the Pascal function sin.

4.2 Use the programs minr and rprec (section 4.3) to obtain estimates of the minimum positive real number and the relative precision of your computer. Change minr to estimate the maximum real number and also maxint for your computer (it is preferable if the output is not sent to a hard-copy device!). Determine the constants from machine documentation and compare with the experimental results.

4.3 Given $x = 1.43541$, $y = 0.0187462$, calculate $x*y$ and x/y as they would be computed on the 4-decimal digit machine described in section 4.4. What are the absolute and relative errors of the results? How many correct decimal places/significant figures can you claim?

4.4 If x and y are two mathematical real numbers, what are the maximum absolute and relative errors in $x*y$ and x/y calculated on a base-β floating-point computer with t-digit precision, double-length accumulator and symmetric rounding? (*Hint:* follow the same steps as in exercise 4.3 and use equation 4.4 where appropriate.) Verify that the result of exercise 4.3 satisfy the error bounds.

4.5 Write a program to exercise the library function roundt. The program should read a set of real numbers and print the values back correctly rounded to 4 significant decimal digits. Test the routine as thoroughly as you can (include a wide range of positive and negative numbers and 0.0 in the data set). Make a copy of roundt and modify the code so that the number of significant digits, t, is a parameter. Why do you think that this was not done in the library routine? Test again with t = 6, 2, −2!

4.6 Modify procedure quad (see exercise 1.3) so that it can compute accurately both roots of a quadratic equation, including cases when $b^2 \gg 4* a*c$. Find an equation for which the old and new versions of the procedure give different answers on your computer (this will be machine dependent). Explain why the results are different and say which you believe. Also try the new procedure on $x^2 = 0$ and explain how this is dealt with in your code.

4.7 The numbers $\bar{x}$, $\bar{y}$ are approximations to two positive real numbers x, y with small errors ϵ_x, ϵ_y respectively. Obtain expressions for the absolute errors in $\bar{x} \ln \bar{y}$ and $\bar{x}^{\bar{y}}$ to first order in ϵ_x, ϵ_y, assuming that exact arithmetic is used in the computation.

The numbers x and y are given as 12.42 and 1.764 rounded to 4 significant figures. Evaluate $x \ln y$ and x^y, and determine how many correct decimal places can be claimed in the answers.

4.8 For the integral

$$P_n = \int_0^1 x^n e^{-\pi x} \, dx$$

show that

$$\pi P_n = n P_{n-1} - e^{-\pi}$$

and

$$0 < P_n < P_{n-1}$$

for $n \geqslant 1$. Devise a stable algorithm to evaluate P_n for $0 \leqslant n \leqslant N$. Program the algorithm and execute with $N = 20$. Check that the numerical values $P_0, P_1, P_2, \ldots$ form a decreasing sequence of positive numbers and that P_0 has the correct value.

4.9 The Bessel functions $J_n(x)$ satisfy the recurrence relation

$$J_{n+1}(x) - (2n/x) J_n(x) + J_{n-1}(x) = 0, \quad n \geqslant 1$$

Write a program to compute and print the numerical values $J_2(x), J_3(x), \ldots, J_N(x)$, given values for x, N, $J_0(x)$ and $J_1(x)$ (you should not use an array to store the sequence!). Given $J_0(1) = 0.7652$, $J_1(1) = 0.4401$ to 4 decimal places, use the program to compute $J_2(1), \ldots, J_{10}(1)$. Do you believe the results? Explain your reasons and give a simple analysis.

4.10 The formula in exercise 3.10

$$f'(x_0) \approx \frac{f(x_0 + h) - f(x_0 - h)}{2h}$$

may be used to approximate the derivative of a function $f(x)$ at x_0 provided f is sufficiently differentiable. Use the formula with a calculator to obtain approximations to the derivative of e^x at $x_0 = 1$, taking $h = 10^{-1}, 10^{-2}, \ldots, 10^{-6}$. Determine the absolute error in each case. Does the error behave as you would expect? Can you explain why the error first decreases, then increases as h $\rightarrow$ 0? (*Hint*: there are two different contributions to the error, truncation error of the approximation formula and rounding error in the function values.)

Part 2

DEVELOPMENT OF MATHEMATICAL SOFTWARE

5 Solution of Non-linear Equations in One Variable: Newton's Method

What I tell you three times is true.

Lewis Carroll, *Hunting of the Snark*

So far we have laid the foundations for our mathematical library. We have seen that the production of mathematical software involves the design of algorithms which are reliable, adaptable and efficient, and the careful coding of these as library routines.

In the next two chapters we shall develop algorithms and library codes for a problem which arises in many practical applications. For a given real-valued function $f(x)$, it is required to determine one or more numbers α such that $f(\alpha) = 0$. The number α, which may be real or complex, is called a *root* of the equation $f(x) = 0$ or a *zero* of the function f.

In most cases the problem of computing roots of an equation must be tackled numerically using techniques of *iteration* (successive approximation). It is rare to be able to solve a class of equations by means of a closed formula (this can be done for quadratics, and it is possible, though usually not desirable, for polynomials up to degree 4). Our intention in these chapters is to develop library routines to compute the real roots of a general algebraic equation $f(x) = 0$. The problem of computing the zeros of a polynomial will be treated as a special case.

5.1 ROOTS OF EQUATIONS

As our intention is to design a general-purpose algorithm, we must try to take account of all possibilities. The equation to be solved may not possess a root, or if a root exists it may not be unique, as the following examples show.

(i) $\ell n(x) = 0$ has the unique root $x = 1$;
(ii) $e^x = 0$ has no root, since $e^x > 0$ for all x;
(iii) $\sin x = 0$ has an infinity of roots: $0, \pm\pi, \pm2\pi, \ldots$.

We can check the existence and uniqueness of real roots by plotting a graph; the mathlib routine graf may be used for this purpose. For example, we can plot the function $f(x) = x - e^{-x}$ between $x = 0$ and $x = 1$.

95

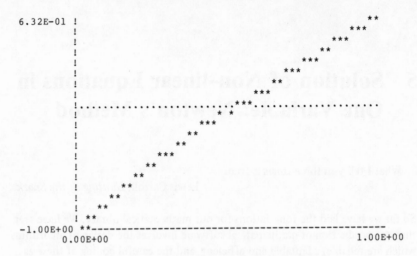

Figure 5.1   Graph of $y = x - e^{-x}$

We see that the equation $x - e^{-x} = 0$ has a root at approximately $x = 0.5$. We can bracket the root by taking, for example, the interval $[0.0, 1.0]$, and we note that the function changes sign over this interval: $f(0.0) = -1.0$, $f(1.0) \approx 0.632$. Roots of this type occur frequently and are called *simple roots*; their general location can be determined graphically or by tabulating the function.

Simple Roots

Definition 5.1

For a function $f(x)$, a *simple root* is a number α such that $f(\alpha) = 0$ and $f'(\alpha) \neq 0$. □

The observation that the function changes sign over an interval containing a simple root suggests the following conditions for existence and uniqueness of roots (these are *sufficient* conditions).

Theorem 5.1

(i)   *Existence*   If a function f is continuous on a closed interval $[a, b]$, and if $f(a)f(b) < 0$, then there is at least one real root of $f(x) = 0$ in (a, b).

(ii)   *Uniqueness*   If in addition f is differentiable and $f'(x)$ has a constant sign in (a, b), then there is exactly one root in (a, b) and this is a simple root.

Proof

(i)   This follows from the intermediate value theorem and is discussed in section 3.2.

(ii)   Suppose that there are two roots α_1, $\alpha_2 \in (a, b)$ with $\alpha_1 \neq \alpha_2$; thus $f(\alpha_1) = f(\alpha_2) = 0$. By the mean value theorem (theorem 3.2) there is some point ξ between α_1 and α_2 at which $f'(\xi) = 0$. This contradicts the condition that $f'(x)$ has a constant sign in (a, b) (that is, $f'(x)$ should be positive or negative for $a < x < b$). Hence $\alpha_1 = \alpha_2$ (= α, say). Furthermore, since $\alpha \in (a, b)$, $f'(\alpha) \neq 0$, and thus α is a simple root. □

As we have seen, the function $f(x) = x - e^{-x}$ changes sign over $[0.0, 1.0]$. Since f is continuous, this guarantees the existence of at least one root in $(0.0, 1.0)$. In addition, $f'(x) = 1 + e^{-x}$, and this is positive for all x. By theorem 5.1 (part (ii)) there is a unique simple root in $(0.0, 1.0)$.

For our purposes the main result of theorem 5.1 is the existence condition. The investigator locates an interval over which the function is continuous and changes sign. With this as a starting point, we can develop a robust algorithm based on bisection (see section 3.2) and incorporate more efficient methods to refine the estimate once we are close to a root.

Finding complex roots is a more difficult problem. These cannot be located graphically, and there is no obvious means of bracketing such roots. Various special methods have been developed to determine real and complex zeros of polynomials (some of these methods are described and analysed by Henrici (1964)). In this book we shall restrict ourselves to the problem of determining real roots.

Multiple Roots

The function

$$f(x) = (x - 1)^2 e^x$$

has a zero at $x = 1$. Clearly, $f(x) \geqslant 0$ for all x and therefore the function does not change sign about $x = 1$. If we differentiate f we find that

$$f'(x) = 2(x - 1)\, e^x + (x - 1)^2 e^x$$

and so $f'(x)$ also has a zero at $x = 1$. We say that the equation $f(x) = 0$ has a *double root* at $x = 1$. This is a special case of the following.

Definition 5.2

For a function $f(x)$ which is sufficiently differentiable, a *root of multiplicity m* is a number α such that

$$f(\alpha) = f'(\alpha) = \ldots = f^{(m-1)}(\alpha) = 0$$

but

$$f^{(m)}(\alpha) \neq 0 \quad □$$

For the function $f(x) = (x - 1)^2 e^x$ we see that $f(1) = 0$, $f'(1) = 0$, but $f''(1) = 2e \neq 0$; hence $x = 1$ is a root of multiplicity 2.

Multiple roots are more difficult to determine than simple roots. One reason, for a root of even multiplicity, is that we are unable to bracket the root by looking for a change of sign in $f(x)$, and therefore we cannot use bracketing methods such as the bisection method. We can avoid this difficulty, if the derivative of f is readily available, by working with the function

$$F(x) = \frac{f(x)}{f'(x)}$$

For example, with $f(x) = (x-1)^2 e^x$, we have

$$F(x) = \frac{(x-1)^2 e^x}{2(x-1)e^x + (x-1)^2 e^x} = \frac{x-1}{x+1}$$

Thus the equation $F(x) = 0$ has a simple root at $x = 1$. It is left as an exercise for the reader to show that if $f(x) = (x - \alpha)^m \Phi(x)$, with $\Phi(\alpha)$ finite and non-zero, then $f(x)$ has a zero of multiplicity m at $x = \alpha$, whereas the function $F(x) = f(x)/f'(x)$ has a simple zero at the same point.

5.2   NEWTON'S METHOD

A root of an equation $f(x) = 0$ is computed in general by a process of iteration. Some methods, such as the bisection method, produce two sequences $\{a_k\}$, $\{b_k\}$, where the interval $[a_k, b_k]$ brackets the root at every step. Other iterative methods produce a single sequence $\{x_k\}$, $k = 0, 1, \ldots$, which under suitable conditions converges to a root of $f(x) = 0$.

Bracketing methods are preferable in a general-purpose algorithm, since the double sequence is guaranteed to converge to a root whenever the function satisfies the existence condition of theorem 5.1. We shall, however, first discuss a method which produces only a single sequence; under certain conditions on f and x_0 this method determines the root in far fewer steps than the bisection method. Later we shall see how to combine the ideas of bracketing and rapid convergence to obtain a general-purpose algorithm suitable for the mathlib library.

If we have an initial approximation x_0 to a root α (obtained, for example, graphically), we can calculate a new approximation to α as follows. In figure 5.2 the tangent to the curve $y = f(x)$ at the point $|x_0, f(x_0)|$ intersects the x axis at x_1, and it follows that

$$f'(x_0) = \tan \theta = \frac{f(x_0)}{x_0 - x_1}$$

This can be solved for x_1 to give

$$x_1 = x_0 - \frac{f(x_0)}{f'(x_0)}$$

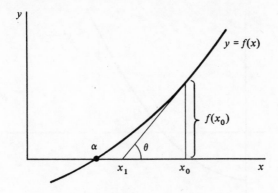

Figure 5.2   One step of Newton's method

Provided that x_0 is sufficiently close to α, and provided that $f'(x_0) \neq 0$, this will give an improved approximation to the root. The process can then be repeated with x_1 to obtain a new approximation x_2, and so on. We define an *iteration*

$$x_{k+1} = x_k - \frac{f(x_k)}{f'(x_k)}, \quad k = 0, 1, \ldots \tag{5.1}$$

which is known as *Newton's method*.

   An important property of Newton's method is the *local convergence property*: if x_0 is chosen 'sufficiently close' to a root α, then the sequence

$$\{x_0, x_1, x_2, \ldots\}$$

converges† to α.

   A standard application of Newton's method is to the calculation of square roots. This shows how well the method can work under favourable conditions. Applying the formula in equation 5.1 to the function

$$f(x) = x^2 - c \quad (\text{where } c > 0)$$

we have

$$x_{k+1} = x_k - \frac{x_k{}^2 - c}{2x_k}$$

$$= \tfrac{1}{2}\left(x_k + \frac{c}{x_k}\right), \quad k = 0, 1, \ldots \tag{5.2}$$

† Sufficient conditions for local convergence are f' non-zero and f'' continuous in an interval containing the root. See Henrici (1964), chapter 4.

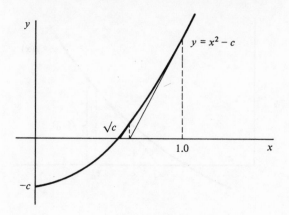

Figure 5.3   Newton's method for $\sqrt{c}$ $(0 < c < 1)$

Example 5.1

Compute $\sqrt{(0.25)}$ by Newton's method, starting with $x_0 = 1$.
We have from formula 5.2, with $c = 0.25$

$$x_1 = \tfrac{1}{2}\left(1 + \frac{0.25}{1}\right) = 0.625$$

$$x_2 = \tfrac{1}{2}\left(0.625 + \frac{0.25}{0.625}\right) = 0.\underline{5}125 \ (1 \text{ figure correct})$$

$$x_3 = \ldots\ldots = 0.\underline{500}\ 152\ 439 \ (3 \text{ figures correct})$$

$$x_4 = \ldots\ldots = 0.\underline{500\ 000}\ 023 \ (7 \text{ figures correct})$$

We note that the number of correct figures increases rapidly, approximately
doubling at each step. This is typical of Newton's method close to a simple root.
The formula in equation 5.2 is a good method of computing a square root on a
hand calculator without a square root key. □

Newton's method, like many iterative methods, is easily programmed for
automatic computation. We simply evaluate the formula 5.1 inside a loop. We
must decide on a stopping criterion for the iteration, and here we rely on the
theory of the method. It can be shown that, provided the estimate x_k is sufficiently close to a simple root α, the error of x_k satisfies

$$|x_k - \alpha| < |x_k - x_{k-1}|$$

The quantity $x_k - x_{k-1}$ is the correction calculated for the latest estimate. We
stop the iteration when the absolute value of the correction is less than a
prescribed tolerance

$$|x_k - x_{k-1}| < \text{tol}$$

Example 5.2

The following fragment of Pascal computes the square root of a positive number c correct to t decimal places.

```
(* c>0 : real ;
   t>0 : integer *)
tol := 0.5 * power i(10.0, -t) ;
x := 1.0 ;  (* arbitrary starting value *)
REPEAT
   xold := x ;                    (* save the latest estimate *)
   x := 0.5 * (xold + c/xold)   (* compute a new estimate *)
UNTIL abs(x-xold) < tol
```

Notes

(i)   It would be a mistake to save all the iterates $x_0, x_1, \ldots, x_k$ in an array; at each stage we require only the values of the two latest estimates.

(ii)   The tolerance is assigned a value before the iteration loop is entered, and the loop is controlled by an absolute error test. Since $tol = 0.5 \times 10^{-t}$, this ensures that the answer is correct to t decimal places (see section 4.2).   □

5.3   IMPLEMENTATION OF NEWTON'S METHOD

We now consider the safeguards which should be incorporated in a general-purpose algorithm to solve non-linear equations in one variable. The discussion will be linked to Newton's method, but the main conclusions will be relevant also to other algorithms to be coded for the mathlib library.

We consider how to program Newton's method to compute a real root of a general equation $f(x) = 0$. The function f and its derivative will be declared as Pascal functions so that they can be changed easily when required. We shall use the test function

$$f(x) = 1 + \frac{3}{x^2} - \frac{4}{x^3}$$

which has a single zero at $x = 1$. From the discussion given so far in this chapter we might write the following program.

```
PROGRAM newton ( input, output ) ;

CONST
   trace = true ;  (* allows intermediate output *)

VAR
   x, correction, tol : real ;

FUNCTION f ( x : real ) : real ;
   BEGIN
   f := 1 + (3 - 4/x) / sqr (x)
   END (* f *) ;
```

```
FUNCTION fd ( x : real ) :  real ;
    BEGIN
    fd := (-6 + 12/x) / (x*sqr(x))
    END (* fd *) ;

BEGIN
read (x, tol) ;  (* input initial estimate and error tolerance *)
REPEAT
    correction := -f(x)/fd(x) ;  (* calculate correction to x
                                      by Newton's method *)
    x := x + correction ;        (* compute new estimate *)
    IF trace THEN
        (* intermediate output required - print the new estimate *)
        writeln (x)
UNTIL abs(correction) < tol ;
writeln (' computed root =', x)
END .
```

A boolean constant trace has been included in the program; with trace = true, intermediate output is produced which assists us to check the performance of the algorithm. If the program is working correctly and such output is not requir-ed, trace can be set to false (the output statement is not deleted as it may be required again, particularly in cases of non-convergence).

In order to develop our program for Newton's method into a form suitable for general use, we must ensure as far as possible that the algorithm is *robust* and *adaptable*.

Convergence and Divergence: Robustness

To explain what is meant by robustness in this context, we consider the output of program newton for two different starting values, $x_0 = 1.3$ and $x_0 = 3.0$ (with tol = 0.5E − 8).

Initial estimate $x_0 = 1.3$

```
6.509285714286E-01
7.933798620472E-01
9.161684790627E-01
9.841444193866E-01
9.993839166239E-01
9.999990518432E-01
9.999999999978E-01
1.000000000000E+00
computed root = 1.000000000000E+00
```

Initial estimate $x_0 = 3.0$

```
1.900000000000E+01
1.306529411765E+03
3.722836164454E+08
8.599447000612E+24
1.059888847309E+74
```

Program fails (floating-point overflow)

In the first case the iterates converge rapidly to the root 1; in the second case the iterates diverge rapidly and the program fails. A runtime failure means that the algorithm is not robust. We seek an explanation for this in the graph of the test function.

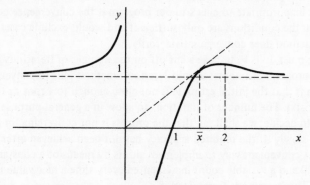

Figure 5.4   Graph of $y = 1 + 3/x^2 - 4/x^3$

It is clear from the graph (and it can be proved) that if x_0 is chosen between 0 and 1, the sequence $\{x_0, x_1, x_2, \ldots\}$ approaches $x = 1$ steadily from the left (Newton's method converges *monotonically* to the root). The reason is that the function has a negative curvature, $f''(x) < 0$, in this interval†; one or two steps of Newton's method drawn in figure 5.4 will indicate the general behaviour. In the output for $x_0 = 1.3$ the first step takes us to $x_1 = 0.6509\ldots$, which lies between 0 and 1, and convergence then proceeds steadily to the root. On the other hand, for $x_0 = 3.0$ the first step takes us away from the root ($x_1 = 19.00\ldots$), and the process then diverges.

The limiting case is given by the point $\bar{x}$ in figure 5.4, where the tangent to the curve passes through the origin. We have

$$f'(\bar{x}) = \frac{f(\bar{x})}{\bar{x} - 0}$$

and by substitution we find that $\bar{x}$ is a root of the cubic equation

$$x^3 + 9x - 16 = 0$$

(in fact, $\bar{x} \approx 1.45$). We conclude that Newton's method will converge to the root 1 for any choice of x_0 satisfying $0 < x_0 < \bar{x}$.‡

†We say that f is *concave* in this interval. The function in example 5.1, $f(x) = x^2 - c$, is *convex*: $f''(x) > 0$ for all x. This accounts for the steady convergence of Newton's square root algorithm to $\sqrt{c}$ from the right of the root.

‡If a function f has a simple root α, then under fairly weak conditions on f we can find an interval I bracketing α such that f'' is either $\geqslant 0$ or $\leqslant 0$ throughout I, and if x_0 lies in I then x_1 produced by Newton's method also lies in I. We are then 'sufficiently close' to the root in the sense of the local convergence property (section 5.2).

The above discussion can be formalised as *non-local convergence theorems* which give sufficient conditions for Newton's method to converge from any starting point in an interval containing a root (see Henrici (1964), chapter 4). These theorems provide a theoretical basis for the use of Newton's method, but it would be inappropriate to check in our program if the convergence conditions are satisfied; the conditions are only sufficient and would exclude cases where Newton's method does converge satisfactorily.

What we must do is to impose a cut-off on the number of iterations allowed, so that the program terminates even if Newton's method fails to converge (this usually means that the initial estimate is not close enough to a root or that no real root exists). The number of iterations to allow in a general-purpose algorithm is difficult to decide; we shall say that the process is not converging, or is converging too slowly, if the required accuracy has not been achieved after 20 iterations. A convenient way to implement this is by means of a constant maxiter = 20 and a variable count incremented every time a new value for x is calculated. The iteration is terminated if

```
count = maxiter
```

Our program is still not robust. In the case of the test function $f(x) = 1 + 3/x^2 - 4/x^3$ we have $f'(2) = 0$, and if the user happens to choose the starting value $x_0 = 2$, Newton's method will fail at once because of zero division. There are two possible actions

(i)   terminate execution and report failure to the user;
(ii)   perturb x and continue.

Since we are already limiting the number of iterations allowed, the second course is quite safe. We replace the statement which computes correction by

```
grad := fd(x) ;
IF abs(grad) < smallreal THEN
   correction := abs(x) + 1.0
ELSE
   correction := -f(x)/grad
```

Thus, when $|f'(x)|$ is zero or very small, x is perturbed by an amount $|x| + 1$. This quantity combines a relative and an absolute perturbation. It is chosen so that the convergence test will not be prematurely satisfied even if $|x|$ is very large or very small.

The Convergence Test: Adaptability

We have gone some way towards adaptability above, using the constant smallreal (100 times the minimum positive real number). We must now make the convergence test adaptable. The new estimate of the root is computed by the statement

```
x := x + correction
```

and the iteration is terminated if

```
abs(correction) < tol
```

The program will be executed on a computer which performs arithmetic to a finite number of digits (finite precision: see chapter 4). Clearly, this will limit the accuracy attainable for the computed root; if tol is too small, the convergence test may never be satisfied.

It is not meaningful to seek an accuracy which is within the possible rounding error incurred in computing the sum x + correction. We recall from section 4.3 that a safe minimum tolerance to use in error tests is given by the library utility function tolmin(x). Thus, to make the convergence test adaptable we write it as

```
abs(correction) < tol + tolmin(x)
```

The expression on the right provides a 'safety net'; the user who wishes to compute a root to full machine accuracy can then safely set tol = 0.0.

Computer Evaluation of $f(x)$: Limiting Precision

There is one further test which should be included in our program. If the current estimate of the root is x and the computed value of $f(x)$ is zero, then we should terminate the iteration since there is no point in continuing. Since 'zero' on the computer corresponds to a finite interval of real numbers (see chapter 4), this situation may be reached before the convergence test is satisfied.

We must now ask what is meant by '$f(x) = 0$' on the computer. The evaluation of a function f involves rounding error and possibly trunction error; thus we cannot evaluate the function exactly. If $\hat{f}(x)$ denotes the computed value of f at x, then we have an error

$$e(x) = \hat{f}(x) - f(x)$$

The equation we actually solve on the computer is $\hat{f}(x) = 0$, and the result obtained by any method is an approximation to a root $\hat{\alpha}$ of $\hat{f}(x) = 0$, instead of the required root α of $f(x) = 0$. It can be shown by the mean value theorem that

$$\alpha - \hat{\alpha} = \frac{e(\hat{\alpha})}{f'(\xi)}$$

where e is the error function defined above and ξ is a point lying between α and $\hat{\alpha}$. Thus the difference $\alpha - \hat{\alpha}$ depends critically on the derivative of f close to the root as well as on the error in the function value. If $|f'(\xi)|$ is small, the function is said to be *ill-conditioned* with respect to α. The conditioning, and hence the attainable accuracy, is likely to be much poorer for a multiple root (with $f'(\alpha) = 0$) than for a simple root.

There is usually little we can do about an ill-conditioned function apart from making sure that $f(x)$ is computed to the best possible accuracy. We do, however, wish to terminate the iteration safely before the computed value $\hat{f}(x)$ *underflows* on the computer (Pascal systems often give a runtime failure on underflow). The third terminating condition is therefore expressed as

```
abs(fx) < smallreal
```

where fx holds the latest computed value f(x). This does not necessarily mean that the estimate x to the root is correct to the required accuracy (although it may be). It does mean that the root $\hat{\alpha}$ of $\hat{f}(x) = 0$ has been computed to the best possible accuracy within the limits of the computer. We therefore refer to this as a *limiting precision test*.

To summarise this section, the iteration loop in Newton's method should be terminated when any of the following conditions is true

```
   (abs(correction) < tol + tolmin(x))  (* required accuracy
                                                   attained *)
OR (count = maxiter)         (* maximum iterations used up *)
OR (abs(fx) < smallreal)     (* limiting precision reached *)
```

As a precaution, a check should also be included for zero derivative

```
abs(grad) < smallreal
```

In the next section we show how to incorporate these refinements in a library code for solving polynomial equations by Newton's method.

5.4   NEWTON'S METHOD FOR POLYNOMIALS: PROCEDURE POLYNEWT

We do not implement Newton's method for a general function in the mathlib library. This is because of the difficulty of making the algorithm really robust, and also because it can be very inconvenient for the user to have to supply the derivative of the function. It may be wondered why we have bothered to discuss Newton's method at all when we can use the bisection method, which possesses guaranteed convergence and does not require the derivative. One reason is that, close to a root, Newton's method does converge rapidly, and this is important when we require roots to a high accuracy. Also, bracketing methods in general are not applicable if we wish to determine complex roots or roots of a system of simultaneous non-linear equations.

In the special case when the function is a polynomial

$$p_n(x) = a_0 + a_1 x + \ldots + a_n x^n$$

the drawbacks of Newton's method are less severe. We recall that p_n and $p_n{}'$ can be evaluated by means of the mathlib routine polval (see section 3.3). Using this routine, and building on the discussion in the last section, we write the following library code to determine a real root of a polynomial equation. (The *deflated polynomial* referred to below is a polynomial $q_{n-1}(x)$, derived from $p_n(x)$ by

Horner's scheme, which can be used to obtain further roots of the equation $p_n(x) = 0$. This will be described more fully after we have given the library code.)

```
PROCEDURE polnewt ( n     : index ;     (* degree of the
                                            polynomial          *)
                    VAR a  : coeff ;     (* array of coefficients
                                            of the polynomial    *)
                    tol    : real  ;     (* absolute error
                                            tolerance           *)
                    trace  : boolean ;   (* when trace=true a trace
                                            of iteration is written
                                            to the output file   *)
                    VAR root : real ;    (* on entry, initial
                                            estimate of the root
                                            supplied by the user;
                                            on exit, final
                                            estimate of the root *)
                    VAR b  : coeff ;     (* on exit, array of
                                            coefficients of the
                                            deflated polynomial  *)
                    VAR eflag : integer  (* error indicator      *)
                  ) ;

(* Newton's method is used to compute a real root of the equation
        p(x) = a[0] + a[1]*x + ... + a[n]*x^n = 0.
The user must supply an initial estimate in parameter root.
The polynomial and its derivative are evaluated by means of
the library routine polval. The iteration terminates normally
when the correction to the computed root is less than
             tol + tolmin(root),
where tolmin is a library utility function which determines
the minimum tolerance to be used in error tests, or when the
limiting precision of the machine is reached, i.e.
             abs(p(root)) < smallreal.
The remaining roots can be determined from the deflated
polynomial
        b[0] + b[1]*x + ... + b[n-1]*x^(n-1)
(the coefficients b[0], ... ,b[n-1] are returned in the
array b).
The error number for eflag is:
   =1   Newton's method fails to converge after
        20 iterations (fatal).
Failure to converge means one of the following:  a real root
does not exist (any roots are complex), or the iteration
is converging slowly to a multiple root, or the initial
estimate supplied is not close enough to a real root.
The nature of the failure can be determined by calling polnewt
with the parameter trace set to true; then at the beginning of
each iteration the current estimate of the root, together with
the value of the polynomial and its derivative, are written to
the output file.
Routines called from the library:
polval, tolmin, errormessage        *)

CONST
    maxiter = 20 ;  (* maximum number of iterations allowed *)

VAR
    x, pval, pgrad, correction : real ;
    count, i                   : integer ;
    limitp                     : boolean ;

BEGIN
x := root ;  (* the variable x holds the latest estimate of
                the root; x is initialised to the user-
                supplied estimate *)
```

```
IF trace THEN
   (* a trace of iteration is required - output the heading *)
   BEGIN
   writeln ;  writeln ('*** trace in polnewt ***') ;
   writeln ('      x                p(x)                p''(x)')
   END ;
count := 0 ;  (* initialise the iteration counter *)

REPEAT  (* iteration loop *)
   polval (n, a, x, pval, pgrad) ;  (* evaluate polynomial and
                                        derivative at x *)
   IF trace THEN
      writeln (x:15, '   ', pval:15, '   ', pgrad:15) ;
   (* Test if p(x)=0 to machine precision:   *)
   limitp := abs(pval) < smallreal ;  (* true if limiting
                                          precision reached *)
   IF NOT limitp THEN  (* calculate new estimate of root *)
      BEGIN
      count := count + 1 ;
      (* Perturb x if p'(x)=0 to machine precision *)
      IF abs(pgrad) < smallreal THEN
         correction := abs(x) + 1.0
      ELSE
         (* Calculate correction to x by Newton's method *)
         correction := -pval/pgrad ;
      x := x + correction
      END
   (* Terminate the iteration when one of the following
      conditions is satisfied:                          *)
UNTIL (abs(correction) < tol + tolmin(x)) (* required accuracy
                                              attained *)
   OR (count = maxiter)  (* maximum iterations used up *)
   OR limitp ;           (* limiting precision reached *)

root := x ;  (* assign the final estimate to root *)
IF (abs(correction) < tol + tolmin(root)) OR limitp THEN
   (* successful call - a root has been found *)
   BEGIN  (* compute deflated polynomial by Horner's scheme *)
   b[n-1] := a[n] ;
   FOR i := n-1 DOWNTO 1 DO
      b[i-1] := b[i]*root + a[i] ;
   eflag := 0  (* set error indicator and exit *)
   END
ELSE
   (* maximum iterations have been used without convergence -
      issue an error message, set error indicator and quit *)
   errormessage ('polnewt ', 1,
      'iteration fails to converge   ', fatal, eflag)

END (* polnewt *)
```

Notes

(i)   The library utility routine errormessage (see section 2.2) is used to deal with cases where Newton's method fails to converge. If polnewt is called with eflag = 0 and a failure occurs, then the following message is written to the output file

```
*** error in polnewt : 1  iteration fails to converge
```

Alternatively, the error message can be suppressed by calling polnewt with eflag $\neq$ 0. In either case eflag is returned with the error number 1. If a real root has been computed successfully, errormessage is not invoked and eflag is returned as 0.

(ii)   A boolean parameter trace is included in the parameter list. If polnewt is called with trace = true, then at the beginning of each step the current estimate of the root, together with the value of the polynomial and its derivative at that point, are written to the output file. This is to assist the user to check the performance of the algorithm, as will be necessary in cases of non-convergence. The direct printing of intermediate output is a rather primitive device; in a more advanced implementation the values would be passed to a user-defined procedure, giving the user full control over how they are processed.

Example 5.3

We have already met one possible application of polnewt in section 5.3. We wished to determine the point $\bar{x}$ in figure 5.4, and we found that this was a root (actually the only real root) of the equation

$$x^3 + 9x - 16 = 0$$

The following program can be used to compute $\bar{x}$.

```
PROGRAM testpolnewt ( input, output ) ;
(* This is an example program to illustrate the use of
   PROCEDURE polnewt to compute a real root of a
   polynomial equation *)

CONST
    tol = 0.0 ;   (* absolute error tolerance - zero value means
                     that full machine accuracy is required  *)
    trace = true ;   (* a trace of the iteration is required *)

TYPE
    index = 0..100 ;
    coeff = ARRAY [index] OF real ;

VAR
    eflag : integer ;
    i, n  : index   ;
    root  : real    ;
    a, b  : coeff   ;

PROCEDURE polnewt ( n : index ;   VAR a : coeff ;   tol : real ;
                    trace : boolean ;   VAR root : real ;
                    VAR b : coeff ;   VAR eflag : integer ) ;
    EXTERN ;

BEGIN
read (n) ;        (* degree of polynomial *)
FOR i := 0 TO n DO
    read (a[i]) ;
read (root) ;    (* initial estimate of root *)
writeln ('example program using polnewt') ;
eflag := 0 ;      (* error messages to output file *)

(* Call polnewt to compute a real root *)
polnewt (n, a, tol, trace, root, b, eflag) ;
IF eflag = 0 THEN
    BEGIN (* a root has been found *)
    writeln ;   writeln ('computed root =', root) ;
    writeln ;
    writeln ('coefficients of the deflated polynomial') ;
    FOR i := 0 TO n-1 DO
        writeln ('b[', i:2, ' ] =', b[i])
    END
END .
```

From figure 5.4 we see that $x_0 = 1.5$ is a good initial estimate. We supply the data

```
     3
  -16.0  9.0  0.0  1.0
     1.5
```

The output is

```
example program using polnewt

*** trace in polnewt ***
      x                  p(x)                p'(x)
  1.4444444E+00      1.3717421E-02      1.5259259E+01
  1.4435455E+00      3.5011445E-06      1.5251471E+01
  1.4435453E+00      2.2737368E-13      1.5251469E+01
  1.4435453E+00     -1.5543122E-15      1.5251469E+01

computed root = 1.443545257674E+00

coefficients of the deflated polynomial
b[ 0 ] = 1.108382291095E+01
b[ 1 ] = 1.443545257674E+00
b[ 2 ] = 1.000000000000E+00
```

The iteration has converged to full machine accuracy (at least 13 correct figures) after 4 or 5 steps of Newton's method. □

The Deflated Polynomial

A polynomial of degree n

$$p_n(x) = a_0 + a_1 x + \ldots + a_n x^n$$

can be written in the factorised form

$$p_n(x) = a_n(x - \alpha_1)(x - \alpha_2) \ldots (x - \alpha_n)$$

where $\alpha_1, \alpha_2, \ldots, \alpha_n$ are the zeros of the polynomial, that is, roots of the equation $p_n(x) = 0$. (It is a fundamental result of algebra that every polynomial of degree n has exactly n zeros, provided that any multiple zeros, corresponding to repeated factors, are counted according to their multiplicity.) The numbers α_i may be real or complex. It can be shown that if the coefficients $a_0, a_1, \ldots, a_n$ are real, then any complex zeros occur in conjugate pairs.

Example 5.4

The polynomial of degree 5

$$p_5(x) = 1 - \frac{5}{2}x + 3x^2 - 3x^3 + 2x^4 - \frac{1}{2}x^5$$

has the factorised form

$$p_5(x) = -\frac{1}{2}(x-1)^2 \ (x-2) \ (x^2+1)$$

Thus the equation $p_5(x) = 0$ has a double root at $x = 1$, a simple root at $x = 2$ and a pair of complex conjugate roots $\pm i$. The graph of the polynomial is shown in figure 5.5.

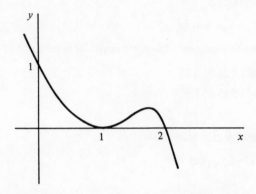

Figure 5.5 Graph of the polynomial $p_5(x)$ □

It is sometimes required to determine all the roots of a polynomial equation. We now consider how we might use polnewt to compute all the real roots. To do this we observe that if we have obtained one real root α, then the polynomial can be factorised as

$$p_n(x) = (x - \alpha) \ q_{n-1}(x)$$

Here, q_{n-1} is a polynomial of degree $n - 1$ called the *deflated polynomial.* (In practice we have only a close estimate of the root, $\bar{x} \approx \alpha$, so that

$$p_n(x) = (x - \bar{x}) \ \hat{q}_{n-1}(x) + r$$

where r is a small constant reminder.) We have seen in equations 3.2 (section 3.3) that the coefficients $b_0, b_1, \ldots, b_{n-1}$ of $\hat{q}_{n-1}$ can be computed by means of Horner's scheme

$$b_{n-1} = a_n$$
$$b_{i-1} = b_i \bar{x} + a_i, \quad i = n - 1, \ldots, 1$$

We now seek a zero of $\hat{q}_{n-1}$; this will be another zero of p_n (approximately, provided the remainder r is small). The process can be repeated to find approximations to all the real zeros of p_n.

Example 5.5

For the cubic equation in example 5.3

$$p_3(x) = x^3 + 9x - 16 = 0$$

we obtained the real root

$$\overline{x} = 1.443\ 545\ 257\ 674$$

The deflated polynomial

$$q_2(x) = b_0 + b_1 x + b_2 x^2$$

was also computed by polnewt; the coefficients were returned as

$$b_0 = 11.083\ 822\ 910\ 95$$
$$b_1 = \ \ 1.443\ 545\ 257\ 674$$
$$b_2 = \ \ 1.0$$

It is easy to check by calculator that

$$p_3(x) = (x - \overline{x})\,q_2(x) + r$$

where $|r| \approx 10^{-9}$

The zeros of the quadratic $q_2(x)$ in this case are complex. If we input the co-efficients b_0, b_1, b_2 to testpolnewt, we find that Newton's method fails to converge from any starting point and eflag is returned as 1. (Newton's method can be used to find complex roots but only with complex initial estimates and complex arithmetic.) □

To compute all the real zeros of a polynomial we use the deflation technique repeatedly. We should be aware, however, that this can lead to a serious growth of error. The sources of error at each stage are

(i)   the error in the coefficients of the deflated polynomial passed on from the previous stage;
(ii)   the error in determining a zero of this polynomial;
(iii) rounding error in computing the new deflated polynomial.

We see from (i) that the error *propagates* from one stage to the next, and this can have a serious effect on the accuracy of the computed results especially for polynomials of high degree. It is therefore important that, having obtained an estimate of a zero from the deflated polynomial, we use Newton's method in the full polynomial to sharpen this estimate. An outline of a suitable algorithm is

```
BEGIN
Input the degree n and read the coefficients
a[0],...,a[n] into an array a ;
Copy the coefficients into an array asave ;
REPEAT
    Use polnewt with the coefficients from a and an
    arbitrary starting value 0.0 to seek a real root ;
```

```
   IF an estimate x of a real root has been found THEN
      BEGIN
      Save the deflated polynomial ;
      If the current degree is less than n, use polnewt
      with the original polynomial from asave to sharpen
      the estimate x ;
      Process (output or store) the final estimate of
      the root ;
      Reduce the degree by 1 and copy the deflated
      polynomial into a
      END
UNTIL n real roots have been found
   OR polnewt fails to converge
END .
```

In certain cases this unsophisticated approach may fail to find some real zeros of a polynomial because of error growth in the deflation process. An estimate of a zero obtained from the deflated polynomial, when refined in the full polynomial, may converge to a different zero; thus some zeros may be found more than once and others not at all. It can be shown that the error growth is minimised if the zeros are extracted in increasing order of absolute value. We may seek to achieve this by using at each stage a more appropriate starting value than 0.0; for an account of one such technique, and applications to finding complex zeros, see Madsen (1973).

Multiple Roots

We have seen that Newton's method, as implemented in polnewt, fails to converge if the initial estimate is not close enough to a real root or if a real root does not exist. There is one further case we must consider. Suppose we use testpolnewt to seek a real root of the equation in example 5.4.

$$1 - \frac{5}{2} x + 3x^2 - 3x^3 + 2x^4 - \frac{1}{2} x^5 = 0$$

which has a double root at $x = 1$ and a simple root at $x = 2$. Starting with $x_0 = 0.0$, the program fails to find any roots. The output is in abbreviated form

```
example program using polnewt

*** trace in polnewt ***
      x                 p(x)              p'(x)
4.0000000E-01     3.3408000E-01     -1.0920000E+00
7.0593407E-01     8.3835326E-02     -5.5596801E-01

. . . . . . . . . . . .     . . . . . . . . . . . .     . . . . . . . . . . . .

9.9998271E-01     2.9904980E-10     -3.4586121E-05
9.9999135E-01     7.4762571E-11     -1.7293069E-05
9.9999568E-01     1.8690646E-11     -8.6465335E-06
9.9999784E-01     4.6723458E-12     -4.3232642E-06
*** error in polnewt : 1   iteration fails to converge
```

We see from the trace that the iteration is converging very slowly to the double root at $x = 1$ (a total of 30 steps would be required to reach limiting precision, and the answer found is 0.999999982. . .). It appears that Newton's method suffers a serious loss of 'converging power' near a multiple root. The reason for this will be investigated in the next section.

5.5   ORDER OF CONVERGENCE

Newton's method (equation 5.1) is a special case of a *one-point iterative method*

$$x_{k+1} = g(x_k), \quad k = 0, 1, \ldots$$

(only one previous estimate is used at each step to compute the next estimate). For Newton's method applied to $f(x) = 0$, we see that $g(x) = x - f(x)/f'(x)$. If the sequence $\{x_0, x_1, \ldots\}$ produced by a one-point method converges to a limit α, then $\alpha = g(\alpha)$†, that is, α is a root of $x - g(x) = 0$.

It is not enough to know that an iterative method converges to a root of an equation. We must also consider the rate of convergence: how many steps are required to determine a root to a given accuracy? We have seen in section 5.4 that this is important; polnewt often fails to converge to a multiple root in the 20 steps allowed. Unfortunately, it is normally not practicable to predict the number of steps which an iterative method will require to determine a root. We can, however, gain some insight into the rate of convergence as follows. First we define the *error* ϵ_k of the estimate x_k,

$$\epsilon_k = x_k - \alpha$$

Method 1

The following fragment of code uses a one-point iterative method to solve the equation $x - e^{-x} = 0$ (the graph of this function is shown in figure 5.1).

```
(* tol = 0.5E-6 *)
x := 0.5 ;  (* initial estimate *)
REPEAT
   xold := x ;
   x := exp(-xold)
UNTIL abs(x-xold) < tol
```

This method is called a simple iteration; we have rearranged the equation into the form $x = e^{-x}$ and derived the iteration

$$x_{k+1} = e^{-x_k}, \quad k = 0, 1, \ldots$$

Method 2

The following code also solves the equation

$$x - e^{-x} = 0.$$

†Provided that g is continuous in an interval containing α.

```
(* tol = 0.5E-6 *)
x := 0.5 ;
REPEAT
    xold := x ;
    x := (xold + 1) / (exp(xold) + 1)
UNTIL abs(x-xold) < tol
```

The iteration used here is obtained directly from Newton's method

$$x_{k+1} = x_k - \frac{x_k - e^{-x_k}}{1 + e^{-x_k}}$$

$$= \frac{x_k + 1}{e^{x_k} + 1}, \quad k = 0, 1, \ldots$$

A simple iteration may or may not converge to a root, depending on the function g and the starting point. Method 1 does converge; it takes 23 steps to determine the root $\alpha = 0.567143.\ldots$. The last few iterates are

k	x_k	ϵ_k	$\epsilon_k/\epsilon_{k-1}$
19	0.567144724	1.434E − 6	−0.567
20	0.567142478	−0.812E − 6	−0.566
21	0.567143751	0.461E − 6	−0.568
22	0.567143029	−0.261E − 6	−0.566
23	0.567143439	0.149E − 6	−0.571

We see that the error ϵ_k is multiplied at each step by a factor which is approximately −0.57, that is, the error is not quite halved in absolute value. This form of convergence, slow and steady, is called *first-order convergence*.

Method 2 determines the root α to the same accuracy in 2 to 3 steps. The results are

k	x_k	ϵ_k	$\epsilon_k/\epsilon_{k-1}$	$\epsilon_k/\epsilon^2_{k-1}$
0	0.<u>5</u>	−67143.3E − 6	—	—
1	0.5<u>66</u>311003	−832.3E − 6	0.0124	−0.185
2	0.567143<u>165</u>	−0.125E − 6	0.0002	−0.180
3	0.567143<u>290</u>	0.000E − 6	—	—

We see that in method 2 the error is approximately −0.18 times the *square* of the preceding error, which means that $\epsilon_k \to 0$ very rapidly ($\epsilon_k/\epsilon_{k-1} \to 0$). This is called *second-order convergence*.

Definition 5.3

An iterative method is said to have *pth order convergence* to a root α if, for a suitable choice of starting value, the method produces a sequence $\{x_k\}$,

$k = 0, 1, \ldots$, converging to α, such that

$$\lim_{k \to \infty} \frac{\epsilon_{k+1}}{\epsilon_k{}^p} = \text{const} \neq 0$$

where

$$\epsilon_k = x_k - \alpha \qquad \square$$

In method 1, $p = 1$ (the method has first-order or *linear* convergence); in method 2, $p = 2$ (the method has second-order or *quadratic* convergence).

We now seek an explanation for the very different behaviour of methods 1 and 2. Both methods determine the root $\alpha = 0.567143\ldots$ of the equation $x - e^{-x} = 0$. In method 1 we have

$$x_{k+1} = e^{-x_k}, \quad k = 0, 1, \ldots$$

Using

$$\epsilon_k = x_k - \alpha$$

we find

$$\begin{aligned} \epsilon_{k+1} + \alpha &= e^{-(\epsilon_k + \alpha)} \\ &= e^{-\alpha} \times e^{-\epsilon_k} \\ &= e^{-\alpha} \left(1 - \epsilon_k + \epsilon_k{}^2/2 - \ldots\right) \end{aligned}$$

Since $\alpha = e^{-\alpha}$, two terms can be cancelled to give

$$\epsilon_{k+1} = \alpha \left(-\epsilon_k + \epsilon_k{}^2/2 - \ldots\right)$$

The iteration is found to be convergent, that is, $\epsilon_k \to 0$ as $k \to \infty$. This means that, for k sufficiently large, $\epsilon_k{}^2$ and higher powers of ϵ_k in the series can be neglected. We then have

$$\epsilon_{k+1} \approx -\alpha \, \epsilon_k$$

The error is thus multiplied by an approximate factor $-\alpha \, (\approx -0.57)$ at each step. We see from the table of results for method 1 that this is in good agreement with the computed results.

Instead of analysing method 2 in a similar way, we consider the general one-point iteration $x_{k+1} = g(x_k)$. The analysis can be unified and at the same time simplified by the use of Taylor's series (see section 3.7). Again with $\epsilon_k = x_k - \alpha$ we have

$$\begin{aligned} \alpha + \epsilon_{k+1} &= g \left(\alpha + \epsilon_k\right) \\ &= g\left(\alpha\right) + \epsilon_k \, g'\left(\alpha\right) + \tfrac{1}{2}\epsilon_k{}^2 g''\left(\alpha\right) + 0\left(\epsilon_k{}^3\right) \end{aligned}$$

provided g is sufficiently differentiable. Cancelling α with $g\left(\alpha\right)$, we obtain the following relation which expresses ϵ_{k+1} in terms of ϵ_k:

$$\epsilon_{k+1} = \epsilon_k g'\left(\alpha\right) + \tfrac{1}{2}\epsilon_k{}^2 g''\left(\alpha\right) + 0\left(\epsilon_k{}^3\right) \tag{5.3}$$

If the iterates converge to α, $\epsilon_k \to 0$ as $k \to \infty$, and from equation 5.3 we then have

$$\lim_{k \to \infty} \frac{\epsilon_{k+1}}{\epsilon_k} = g'(\alpha)$$

In method 1, $g(x) = e^{-x}$, $g'(x) = -e^{-x}$ and $g'(\alpha) = -\alpha \approx -0.57$, confirming our earlier result.

Clearly, for convergence of a one-point iteration to α it is necessary that $|g'(\alpha)| < 1$. It can be shown that a sufficient condition for convergence is $|g'(x)| < 1$ in an interval I containing a root, provided x_0 is chosen close enough to the root so that all the iterates lie in I. (For a more detailed discussion see, for example, Phillips and Taylor (1973), chapter 7, or Henrici (1964), chapter 4.) It can be shown further that if $|g'(x)| \leqslant \frac{1}{2}$ throughout I, then

$$|\epsilon_k| \leqslant |x_k - x_{k-1}|$$

In this case we are justified in taking the absolute value of the correction $x_k - x_{k-1}$ as a bound on the absolute error.

The smaller the value $|g'(\alpha)|$, the faster the iteration will converge close to the root. Ideally we would like $g'(\alpha) = 0$. Then from equation 5.3

$$\epsilon_{k+1} \approx \frac{1}{2} \epsilon_k^2 g''(\alpha)$$

and the method will have second-order convergence. In method 2 above

$$g(x) = \frac{x+1}{e^x + 1}, \quad g'(x) = \frac{e^{-x} - x}{e^x + 2 + e^{-x}}$$

and since α is a root of $x - e^{-x} = 0$, $g'(\alpha) = 0$. Thus method 2 is (at least) second order. A further differentiation shows that $\frac{1}{2} g''(\alpha) = -0.1809$, in agreement with the table of results.

Newton's method in general has second-order convergence to a *simple* root. We have

$$g(x) = x - \frac{f(x)}{f'(x)}, \quad g'(x) = \frac{f(x) f''(x)}{(f'(x))^2}$$

At a simple root $f(\alpha) = 0$, $f'(\alpha) \neq 0$; hence $g'(\alpha) = 0$, and from equation 5.3 Newton's method is at least second order. A further differentiation shows that it is exactly second order if $f''(\alpha) \neq 0$. For a root of multiplicity m, however, (where $f'(\alpha) = \ldots = f^{(m-1)}(\alpha) = 0$), it can be shown that $g'(\alpha) = 1 - 1/m$. Thus for $m > 1$, $g'(\alpha) \neq 0$; Newton's method has only first-order convergence to a multiple root. This accounts for the slow convergence of polnewt observed at the end of section 5.4.

EXERCISES

5.1 Devise an iterative formula based on Newton's method to find the nth root of a real number $c > 0$. Hence, using a calculator, determine $e^{1/3}$ and $\pi^{1/5}$ correct to 5 significant figures.

5.2 Derive Newton's method to approximate a root α of an equation $f(x) = 0$ by expanding $f(\alpha)$ in Taylor's series about x_k, truncating at a suitable term and solving for α. Obtain a higher-order method by including the next term in the Taylor's series. Use a formula based on this method to repeat the calculation in exercise 5.1.

5.3 Prove that if $f(x) = (x - \alpha)^m \Phi(x)$, where $m \geq 1$, $\Phi(\alpha) \neq 0$ and Φ is arbitrarily differentiable, then the equation $f(x) = 0$ has a root of multiplicity m at $x = \alpha$ (see definition 5.2). Show that, on the other hand, the equation $f(x)/f'(x) = 0$ has a simple root at $x = \alpha$.

5.4 Consider the iterative methods of the form

$$x_{k+1} = g(x_k), \quad k = 0, 1, \ldots$$

with

(i) $g(x) = \dfrac{1 + x^2}{c + x}$, $\quad c \neq 0$

(ii) $g(x) = x(2 - cx)$, $\quad c \neq 0$

(iii) $g(x) = \frac{1}{8}(3x + 6c/x - c^2/x^3)$, $\quad c > 0$

What do these formulae compute if they converge? Show that (i) is first order, (ii) is second order and (iii) is third order. Use one of these methods to calculate $1/\pi$ and $1/e$ correct to 5 significant figures without performing a division.

5.5 Use procedure polnewt to compute separately all the zeros of $2x^3 - 4x^2 - 3x + 4$ correct to 6 decimal places (see also exercise 2.4). Repeat for the polynomial $x^4 + x^3 - 7x^2 - x + 5$.

5.6 One equation of state relating the pressure P, molar volume V and absolute temperature T of a gas is the Beattie–Bridgeman equation

$$P = \frac{RT}{V} + \frac{\beta}{V^2} + \frac{\gamma}{V^3} + \frac{\delta}{V^4}$$

where R is the universal gas constant. The parameters β, γ and δ are defined by

$$\beta = RTb - a - \frac{Re}{T^2}$$

$$\gamma = -RTbd + ac - \frac{Rbe}{T^2}$$

$$\delta = \frac{Rbde}{T^2}$$

where a, b, c, d and e are constants for each gas.

Write a program to compute the molar volume V correct to n significant figures, given values for P and T and the appropriate constants. (Use procedure polnewt, obtaining a starting value from the ideal gas law $PV = RT$.) The constants for methane are $a = 2.2769$, $b = 0.05587$, $c = 0.01855$, $d = -0.01587$, $e = 128300$. Compute V in this case to 4 significant figures for $T = 300$ and $P = 20, 50, 100$, given $R = 0.08205$.

5.7 Extend the program in exercise 5.6 to plot a graph of the *compressibility factor, z = PV/RT*, as a function of pressure when T is held constant (use the mathlib procedure graf). The compressibility factor shows how the behaviour of a real gas deviates from that of an ideal gas. Plot three graphs of z against P for methane at $T = 300, 400$ and 500 for $P = 10, 20, \ldots, 200$.

5.8 Write a procedure which calls procedure polnewt to compute, as far as possible, all the real zeros of a polynomial. Use the algorithm for the deflation process outlined in section 5.4. The procedure should return an array containing the zeros and an integer parameter to indicate the number of zeros found. Test on the polynomials of exercise 5.5 and also on the polynomial (Conte and de Boor (1980))

$$x^5 - 15.5x^4 + 77.5x^3 - 155x^2 + 124x - 32$$

5.9 Derive the conditions stated in section 5.5 under which Newton's method has exactly second-order convergence to a root. If $f(x) = (x - \alpha)^m \Phi(x)$, where $m > 1$ and Φ is twice differentiable, show that Newton's method has only first-order convergence to α. If in addition $\Phi(\alpha) \neq 0$, find a constant r such that the following method is at least second order

$$x_{k+1} = x_k - rf(x_k)/f'(x_k)$$

5.10 Write a procedure on the lines of polnewt which can determine efficiently any real zero of $-0.5x^5 + 2x^4 - 3x^3 + 3x^2 - 2.5x + 1$ (example 5.4). Test the procedure with starting values $x_0 = 0$ and $x_0 = 3$. Would this be suitable as a library routine?

6 Solution of Non-linear Equations in One Variable: Interval Methods

If you knows of a better 'ole, go to it.

C. B. Bairnsfather, *Fragments from France*

As we have seen in chapter 5, Newton's method often behaves erratically at a distance from a root. For a general function, if the starting point is not sufficiently close to a root, the sequence of iterates may diverge.

A class of algorithms known as *interval methods*, much more robust than Newton's method alone, can be developed from the bisection process described in section 3.2. The only requirement is a starting interval $[a, b]$ on which the function is continuous and changes sign, that is, $f(a)f(b) \leq 0$. This condition can be tested before the process is started. The general step consists in narrowing the interval, the reduced interval being chosen so that the root is bracketed at every stage.

In this chapter we shall develop and test library codes based on three interval methods, with refinements progressively added to improve the efficiency. The methods are: simple bisection, bisection/secant and bisection/rational interpolation.

6.1 INTERVAL CODE: PROCEDURE BISECT

With a suitable starting interval $[a, b]$ and with proper safeguards, the bisection method converges to a root of $f(x) = 0$ in $[a, b]$. An outline of a suitable algorithm is

```
BEGIN   (* bisection algorithm *)
Check that the function f changes sign over [a, b] ;
WHILE   required accuracy not attained
   AND  limiting precision not reached   DO
   BEGIN
   calculate a new estimate x at the mid-point of [a, b] ;
   evaluate f(x) ;
   use x to halve the interval [a, b] bracketing the root
   END ;
Return the end-point of the final interval at which f is
smaller in absolute value
END .
```

The bisection process terminates in one of two ways:

(i)   The width of the current interval is less than a prescribed tolerance; the process has then converged to the required accuracy.

(ii)   Limiting precision is reached, that is, f(x) = 0 to machine precision; the latest value x should then be returned with a warning.

In the code which follows, the two possible terminating states are identified by boolean variables converged and limitp. The respective conditions are tested immediately on entry to the while loop (it is possible that the initial interval already falls within the prescribed tolerance or that limiting precision applies at a or b).

The code for the bisection algorithm is now given. As usual, we must ensure that convergence tests etc. are adaptable.

```
PROCEDURE bisect (
          FUNCTION f(x:real) : real ;   (* specification of
                                          the function f  *)
                a, b : real ;   (* user-supplied interval
                                  [a, b] such that
                                  f(a)*f(b) <= 0  *)
                tol  : real ;   (* absolute error
                                  tolerance          *)
            VAR root : real ;   (* final estimate
                                  of the root      *)
            VAR eflag : integer (* error indicator *)
                      ) ;

(* The simple bisection method is used to compute a root of the
   equation f(x)=0 in the interval [a, b], where f(a)*f(b) <= 0.
   The function f must be defined and continuous on [a, b].
   The iteration terminates normally when the error in the
   computed root is less than
                  tol + tolmin(root),
   where tolmin is a library utility function which determines
   the minimum tolerance to be used in error tests.
   The error numbers for eflag are:
   =1  The function does not change sign over [a, b],
       i.e.  f(a)*f(b) > 0  (fatal)
   =2  Limiting precision is reached,
       i.e.  abs(f(root)) < smallreal,
       before the convergence test is satisfied. The latest
       estimate is returned in root  (warning).
   Routines called from the library:
   sign, errormessage, tolmin        *)

VAR
   x, fx, fa, fb    : real ;
   converged, limitp : boolean ;

BEGIN
   (* The basis of this routine is, after checking that f changes
      sign over [a,b], to halve the interval [a,b] in stages and
      bracket the root at every stage *)

   (* Check that the function f changes sign over [a, b] -
      note it is not assumed that a <= b. The sign function
      is used to avoid overflow or underflow in the product *)
   fa := f(a) ;   fb := f(b) ;
   IF sign(fa)*fb > 0.0 THEN
       (* issue an error message, set error indicator and quit *)
       errormessage ('bisect ', 1,
           'function does not change sign ', fatal, eflag)
```

```
    ELSE
        BEGIN  (* bisection method to determine a root in [a, b] *)
        (* Test if supplied interval [a, b] lies within specified
           tolerance or if limiting precision applies at a or b *)
        converged := abs(b-a) < tol + tolmin(a) ;
        limitp := (abs(fa) < smallreal) OR (abs(fb) < smallreal) ;

        WHILE  NOT converged  (* required accuracy not attained *)
           AND  NOT limitp (* limiting precision not reached *)  DO
           BEGIN  (* iteration loop *)
           (* Calculate a new estimate of the root
              at the mid-point of [a, b] *)
           x := a + 0.5*(b-a) ;    fx := f(x) ;
           (* Halve the interval containing the root *)
           IF sign(fa)*fx < 0.0 THEN
              BEGIN  (* root lies between a and x *)
              b := x ;   fb := fx
              END
           ELSE
              BEGIN  (* root lies between x and b *)
              a := x ;   fa := fx ;
              END ;
           (* Re-evaluate the control variables *)
           converged := abs(b-a) < tol + tolmin(x) ;
           limitp := abs(fx) < smallreal
           END  (* iteration loop *) ;

        (* Assign the final estimate to root *)
        IF abs(fa) < abs(fb) THEN
           root := a
        ELSE
           root := b ;
        (* Set error indicator, issue any warning and exit *)
        IF converged THEN
           eflag := 0  (* successful call *)
        ELSE
           errormessage ('bisect  ', 2,
              'limiting precision reached   ', warning, eflag)
        END  (* bisection method *)

    END (* bisect *)
```

Notes

(i)   The function f(x) is a parameter to the procedure. The syntax for this depends on the Pascal implementation (see appendix D).

(ii)   The sign function is used in the form sign (fa) when products such as f(a)*f(b) are tested. This is to avoid overflow or underflow which can occur if f(a)*f(b) is computed as it stands.

(iii) ₁ The new estimate x (mid-point of [a, b]) is computed by adding a correction on to a,

$$x: = a + 0.5* (b - a)$$

instead of

$$x: = (a + b)/2$$

In the latter case, rounding error could cause x to lie outside the interval [a, b] when a and b are very close.

(iv) Since the bisection process is guaranteed to terminate, there is no need for a count or cut-off maxiter, and a trace as in polnewt is unnecessary except for illustrative purposes.

In order to employ procedure bisect, the user must write a program in which a specific function f(x) is declared and the end-points a, b of the initial interval and the tolerance tol are specified. As an example, we use the test function discussed in section 5.3

$$f(x) = 1 + 3/x^2 - 4/x^3$$

```
PROGRAM testbisect ( input, output ) ;
(* This is an example program to illustrate the use of an
   interval code from the library to determine a root of
   a general equation f(x)=0 in an interval [a, b].
   The test function used is
        f(x) = 1 + 3/x^2 - 4/x^3
   and the interval code is PROCEDURE bisect *)

CONST
   tol = 0.5E-8 ;

VAR
   a, b, root : real ;
   eflag      : integer ;

(* Declare the library routine *)
PROCEDURE bisect ( FUNCTION f(x:real) : real ;  a, b : real ;
                   tol : real ;  VAR root : real ;
                   VAR eflag : integer ) ;
   EXTERN ;

(* Specify the test function *)
FUNCTION f ( x : real ) :  real ;
   BEGIN
   f := 1 + (3 - 4/x) / sqr(x)
   END (* f *) ;

BEGIN
read (a, b) ;  (* input the bounds on the root *)
writeln ('example program using bisect') ;
writeln ;  writeln ('initial interval ', a, ' ', b) ;
eflag := 0 ;  (* error messages and warnings to output file *)

bisect (f, a, b, tol, root, eflag) ;  (* use bisect to compute a
                                          zero of f(x) in [a,b] *)
IF eflag <> 1 THEN
   BEGIN  (* a root has been found *)
   writeln ;  writeln ('computed root =', root)
   END
END .
```

Example 6.1

Solve the equation $1 + 3/x^2 - 4/x^3 = 0$ correct to 8 decimal places using procedure bisect.

We choose the initial interval [0.5, 5.0] and supply the following data to testbisect:

0.5 5.0

A total of 30 steps is required for convergence. A modified version of bisect is used to output the end-points of the current interval and show the narrowing of the interval at every step (the notation is similar to that in example 3.3). This is for illustration only and is unnecessary in a library routine. The output is in abbreviated form

```
example program using bisect

initial interval  5.000000000000E-01    5.000000000000E+00

*** trace in bisect ***
        5.000000000000E-01      2.750000000000E+00   << b
        5.000000000000E-01      1.625000000000E+00   << b
        5.000000000000E-01      1.062500000000E+00   << b
  b >>  7.812500000000E-01      1.062500000000E+00

        .................       .................

  b >>  9.999999869615E-01      1.000000003725E+00
  b >>  9.999999953434E-01      1.000000003725E+00
  b >>  9.999999995343E-01      1.000000003725E+00

computed root = 9.999999995343E-01
```

If the initial interval $[2.0, 3.0]$ is chosen instead, the output is

```
example program using bisect

initial interval  2.000000000000E+00    3.000000000000E+00
*** error in bisect  : 1   function does not change sign
```

The Bisection/Newton Method

The bisection algorithm is robust. The advantage of robustness in a library routine is evident from example 6.1; we can start at a distance from a root, and (provided the function is continuous) we can be sure of obtaining either an estimate of a root or an informative error message†. In such cases Newton's method would simply diverge. Close to a root, however, the bisection method is very much slower than Newton's method (see the results for $x_0 = 1.3$ in section 5.3).

Our task now is to devise an algorithm which combines the advantages of both methods, that is, robustness and rapid convergence close to a root. An outline of an algorithm which combines bisection and Newton's method is as follows. This has the same structure as the bisection algorithm and differs only in the details of the while loop where x is updated. An initial interval [a, b] bracketing the root is required as before.

```
(* Bisection/Newton algorithm *)
Initialise x to (a+b)/2 ;
WHILE  required accuracy not attained
  AND  limiting precision not reached  DO
```

†Even if the function is only piecewise continuous, the bisection method can be used to locate a discontinuity involving a change of sign.

```
BEGIN
IF f'(x) <> 0 THEN
    calculate a new estimate by Newton's method:
    x := x - f(x)/f'(x) ;
IF x now lies outside (a, b) OR Newton's method is
inapplicable (zero derivative) THEN
    calculate a new estimate by the bisection method:
    x := a + 0.5*(b-a) ;
Evaluate f(x) and use x to narrow the interval [a, b]
as in the bisection algorithm
END
```

The new estimate x from Newton's method is accepted if and only if $a < x < b$, where [a, b] is the current interval. This will generally be the case close to a root. The role of the bisection method is to refine the interval until Newton's method can take over and give final rapid convergence. The root is bracketed at every stage. There is, however, one major drawback to the bisection/Newton algorithm. A derivative function f' must be supplied, and this has to be evaluated at every step even if the estimate from Newton's method is then rejected. Before we write an improved interval code, therefore, we consider some possible alternatives to Newton's method which do not require the derivative.

6.2   ALTERNATIVES TO NEWTON'S METHOD

The requirement of the derivative in Newton's method is a serious disadvantage in the context of a library routine for solving general non-linear equations. It may be difficult or impossible for the user to supply a derivative function, and even if a function can be supplied it is often computationally expensive to evaluate. In this section we examine some alternative methods which do not require the derivative. These may be regarded as modifications of Newton's method,

$$x_{k+1} = x_k - \frac{f(x_k)}{f'(x_k)}, \quad k = 0, 1, \ldots$$

The most useful variant to be considered, the *secant method*, will be incorporated later into an improved library code.

Constant Gradient

The simplest possibility is to replace $f'(x_k)$ by a constant value m, intended to approximate the derivative near the root. Thus we obtain

$$x_{k+1} = x_k - \frac{f(x_k)}{m}, \quad k = 0, 1, \ldots$$

The resulting process is in general first-order. A poor choice of m or x_0 may lead to divergence, but a good choice can give quite rapid convergence. This idea is particularly useful for solving systems of non-linear equations in several variables (see Dahlquist and Björck (1974), chapter 6).

Steffensen's Method

Another possibility is to replace $f'(x_k)$ by an approximating expression dependent on x_k. The replacement

$$f'(x_k) \approx \frac{f(x_k + f(x_k)) - f(x_k)}{f(x_k)}$$

may be motivated by Taylor's series expansion about x_k. Writing f_k for $f(x_k)$, we expand $f(x_k + f_k)$ as

$$f(x_k + f_k) = f(x_k) + f_k f'(x_k) + 0\,(f_k^2)$$

Substituting into the expression above, we find

$$\frac{f(x_k + f(x_k)) - f(x_k)}{f(x_k)} = f'(x_k) + 0\,(f_k)$$

where, as $x_k \to \alpha$, $f_k \to f(\alpha) = 0$. With this replacement for $f'(x_k)$ in Newton's method, we obtain

$$x_{k+1} = x_k - \frac{[f(x_k)]^2}{f(x_k + f(x_k)) - f(x_k)}, \quad k = 0, 1, \ldots$$

This is known as *Steffensen's method*. The method has second-order convergence and behaves in many respects like Newton's method. The derivative, however, is not required; instead, two evaluations of f are necessary at every step.

The Secant Method

The modification of Newton's method which we shall find most useful is obtained by means of the replacement

$$f'(x_k) \approx \frac{f(x_k) - f(x_{k-1})}{x_k - x_{k-1}}$$

This can be established by Taylor's series expansion about x_k or motivated graphically from a diagram.

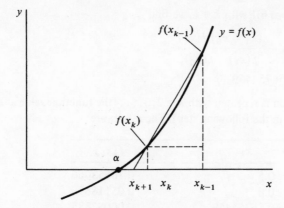

Figure 6.1   One step of the secant method

Figure 6.1 should be compared with figure 5.2 for Newton's method. The tangent in Newton's method is replaced by the secant passing through the two preceding points on the curve. The intersection of the secant with the x-axis is taken as the next estimate x_{k+1}. We express this as

$$
\left.
\begin{aligned}
grad &= \frac{f(x_k) - f(x_{k-1})}{x_k - x_{k-1}} \,, \\[2mm]
x_{k+1} &= x_k - \frac{f(x_k)}{grad} \,, \quad k = 1, 2, \ldots
\end{aligned}
\right\}
\tag{6.1}
$$

(provided $grad \neq 0$)

This is known as the *secant method*.

The secant method differs from the one-point methods considered so far in that it requires a memory of the preceding point x_{k-1} and the function value $f(x_{k-1})$. Initially, two points x_0, x_1 and two function values $f(x_0), f(x_1)$ are required. Thereafter, the computation can be organised so that only *one* evaluation of f is necessary at each step. This is important and is a major advantage of the secant method; function evaluations are usually by far the most expensive part of an iterative computation.

Example 6.2

Solve the equation $x - e^{-x} = 0$ by the secant method correct to 6 decimal places, taking $x_0 = 0.5, x_1 = 0.6$.

We have

$$
f(x) = x - e^{-x}
$$

thus

$$
f(x_0) = -0.10653066
$$
$$
f(x_1) = 0.05118836
$$

Using equations 6.1 with $k = 1$, we find

$$x_2 = x_1 - \frac{(x_1 - x_0)\, f(x_1)}{f(x_1) - f(x_0)}$$

$$= 0.56754459$$

The calculation is repeated with $k = 2, 3, \ldots$ (the function value at each step is saved for use in the following step). The results are

k	x_k	$f(x_k)$
0	0.5	−0.10653066
1	0.6	0.05118836
2	0.56754459	0.00062885
3	0.56714092	−0.00000371
4	0.56714329	$\approx - 10^{-9}$

The computed root is 0.567143 correct to 6 decimal places. This may be compared with the solution of the same problem by simple iteration and Newton's method in section 5.5. □

The secant method behaves in a similar way to Newton's method; in particular, it suffers from the same erratic behaviour at a distance from a root. It can be shown that the order of convergence is

$$\tfrac{1}{2}(1 + \sqrt{5}) = 1.618\ldots$$

Thus, convergence close to a root is a little slower than with Newton's method or Steffensen's method, but this is usually more than offset by the saving in function evaluations at every step after the first.

6.3   IMPROVED INTERVAL CODE: PROCEDURE BISECANT

We now combine the bisection and secant methods to produce an interval iterative algorithm with the following properties:

(i)   Guaranteed convergence to a root subject to the existence condition of theorem 5.1;

(ii)   Rapid (superlinear) convergence close to a simple root;

(iii)   No derivative requirement − only one function evaluation per step.

In the secant method of equations 6.1 the derivative $f'(x_k)$ is approximated by the gradient of the secant at two points, $(x_k, f(x_k))$ and $(x_{k-1}, f(x_{k-1}))$. It follows that two variables are required: x, to hold the current estimate of the root, and xmem, to hold the previous estimate. These are initialised to the endpoints of the user-supplied interval [a, b] . The values are then updated in an iteration loop: the current estimate x is saved in xmem, and a new estimate is

computed. The new estimate x is obtained from the secant method provided
a < x < b, where [a, b] is the current interval bracketing the root; otherwise x
is obtained from the bisection method. The new x and f(x) are then used to
narrow the interval [a, b] as in the bisection algorithm.

```
BEGIN   (* bisection/secant algorithm *)
Check that the function changes sign over [a, b] ;
Initialise two starting points for the secant method:
   x := a,   xmem := b ;
WHILE  required accuracy not attained
   AND  limiting precision not reached   DO
   BEGIN
   Approximate f'(x) by the gradient of the secant
   at x and xmem:
      grad := (f(x)-f(xmem)) / (x-xmem) ;
   Save the current estimate of the root:
      xmem := x ;
   IF grad <> 0 THEN
      calculate a new estimate by the secant method:
      x := x - f(x)/grad ;
   IF x does not lie in (a, b) OR the secant method is
   inapplicable (zero gradient) THEN
      calculate a new estimate by the bisection method:
      x := a + 0.5*(b-a) ;
   Evaluate f(x) and use x to narrow the interval [a, b]
   END ;
Return the end-point of the final interval at which f is
smaller in absolute value
END .
```

A further refinement may now be described. It is possible in certain cases for
the secant method to converge monotonically from one side of the root while
the other end-point of the interval remains fixed.

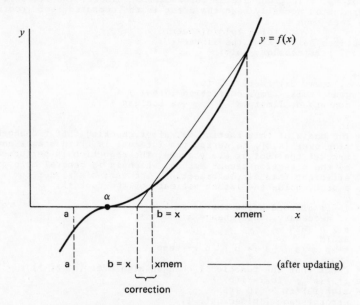

Figure 6.2   Monotonic convergence of the secant method

Since the interval width b — a may be greater than the prescribed tolerance
(tolx = tol + tolmin(x)), the convergence test in our algorithm may not be
satisfied at this stage. The bisection method would then take over, leading to
much slower convergence. We therefore perturb x in such cases as follows.
Denoting the correction to x due to a secant step by correction: = —fx/grad
(see figure 6.2), we test

```
IF abs(correction) < tolx THEN
   (* the secant method has converged *)
   correction := sign(correction)*tolx ;
x := x + correction
```

Thus x is perturbed by an amount tolx in the same direction as the secant step.
The intention is to shift x to the other side of the root. When the interval [a, b]
is next updated, a will then be moved up to x and the interval will be reduced to
a width of tolx. The perturbation also ensures that after completion of a step
x — xmem is never zero to machine precision.

The code for the bisection/secant algorithm is now given; this should be
compared with procedure bisect.

```
PROCEDURE bisecant ( FUNCTION f(x:real) : real ;
                              a, b    : real ;
                              tol     : real ;
                          VAR root    : real ;
                          VAR eflag   : integer ) ;

(* A combination of the bisection and secant iterative methods
   is used to compute a root of the equation f(x)=0 in the
   interval [a, b] , where f(a)*f(b) <= 0. The parameters and
   user specifications are as for PROCEDURE bisect. The iteration
   terminates normally when the error in the computed root (root)
   is less than
                       tol + tolmin(root).
   Routines called from the library:
   sign, errormessage, tolmin        *)

   VAR
       x, xmem, fx, fmem, fa, fb,
       grad, tolx, temp, correction : real ;
       converged, limitp            : boolean ;

   BEGIN
   (* The basis of this routine is, after checking that f changes
      sign over [a,b], to narrow the interval [a,b] in stages and
      bracket the root at every stage. The reduction is performed
      by the bisection/secant method, which is in general more
      efficient than simple bisection. The variable x, set to
      a or b, holds the latest estimate of the root *)

   (* It is assumed below that a <= b;
      if necessary, interchange a and b *)
   IF a > b THEN
       BEGIN
       temp := a ;  a := b ;  b := temp
       END ;
   (* Check that the function f changes sign over [a, b] *)
   fa := f(a) ;   fb := f(b) ;
   IF sign(fa)*fb > 0.0 THEN
       errormessage ('bisecant', 1,
           'function does not change sign ', fatal, eflag)
   ELSE
```

```
      BEGIN   (* bisection/secant method
                 to determine a root in [a, b] *)
      x := a ;   xmem := b ;       (* initialise two points *)
      fx := fa ;   fmem := fb ;   (* for the secant method *)
      tolx := tol + tolmin(x) ;
      (* Test initial interval [a,b] as in bisection algorithm *)
      converged := b-a < tolx ;
      limitp := (abs(fa) < smallreal) OR (abs(fb) < smallreal) ;

      WHILE  NOT converged  (* required accuracy not attained *)
        AND  NOT limitp  (* limiting precision not reached *)  DO
        BEGIN   (* iteration loop *)
        (* Approximate f'(x) by the gradient of the secant
           at x and xmem *)
        grad := (fx/fmem - 1.0) * (fmem/(x-xmem)) ;
        xmem := x ;   fmem := fx ;   (* save current estimate
                                        of the root *)
        (* If the gradient is nonzero to machine
           precision, calculate a new estimate of the
           root by the secant method *)
        IF abs(grad) > smallreal THEN
           BEGIN
           correction := -fx/grad ;
           IF abs(correction) < tolx THEN
              (* the secant method has converged *)
              correction := sign(correction)*tolx ;
           x := x + correction
           END ;
        (* If the estimate does not lie in (a, b) or the
           secant method is inapplicable (zero gradient),
           use the bisection method *)
        IF (x <= a) OR (x >= b) THEN
           x := a + 0.5*(b-a) ;
        (* x is the new estimate of the root *)
        fx := f(x) ;

        (* Narrow the interval containing the root *)
        IF sign(fa)*fx < 0.0 THEN
           BEGIN   (* root lies in [a,x] *)
           b := x ;   fb := fx ;
           END
        ELSE
           BEGIN   (* root lies in [x,b] *)
           a := x ;   fa := fx
           END ;
        (* Re-evaluate the tolerance and control variables *)
        tolx := tol + tolmin(x) ;
        converged := b-a < tolx ;
        limitp := abs(fx) < smallreal
        END   (* iteration loop *) ;

      (* Assign the final estimate to root *)
      IF abs(fa) < abs(fb) THEN
         root := a
      ELSE
         root := b ;
      (* Set error indicator, issue any warning and exit *)
      IF converged THEN
         eflag := 0   (* successful call *)
      ELSE
         errormessage ('bisecant', 2,
            'limiting precision reached    ', warning, eflag)
      END   (* bisection/secant method *)

END (* bisecant *)
```

Notes

(i)   In bisecant, unlike bisect, it is assumed internally for convenience of coding that a $\leqslant$ b. The initial values supplied for a and b are checked and interchanged if necessary.

(ii)   Only one function evaluation f(x) is performed at each step of the iteration. A memory of the function value fmem is kept as well as a memory of the previous iterate xmem.

(iii)   The formula for grad is coded in a form which avoids underflow failure. Underflow could occur in computing the difference fx − fmem, and this would lead to runtime failure on most Pascal systems (and, in any case, inaccurate results).

(iv)   The new estimate x obtained from the secant method is accepted if and only if x lies within the current interval (a, b). This is the purpose of the test

```
IF (x <= a) OR (x >= b) THEN
   (* use bisection method *)
```

Since at the start of each step x is equal to either a or b, this test also detects if x has not been updated by the secant method (because of zero gradient).

The parameter list of procedure **bisecant** is the same as that of procedure **bisect**. We can therefore use the example program **testbisect** given in section 6.1, with the substitution of bisecant for bisect throughout.

Example 6.3

Solve the equation $1 + 3/x^2 - 4/x^3 = 0$ correct to 8 decimal places using procedure **bisecant**.

We use a modified version of bisecant to output the end-points of the current interval and indicate whether a secant step or a bisection step has been taken. The notation is
b − bisection; s − secant; * − special perturbation step
For the initial interval [0.5, 5.0], 10 steps are required for convergence. The output is

```
example program using bisecant

initial interval   5.000000000000E-01    5.000000000000E+00

*** trace in bisecant ***
         5.000000000000E-01    4.756272401434E+00   << s
         5.000000000000E-01    4.524255483729E+00   << s
         5.000000000000E-01    2.512127741865E+00   << b
         5.000000000000E-01    1.506063370932E+00   << b
         5.000000000000E-01    1.003031935466E+00   << b
s >>     9.950205588597E-01    1.003031935466E+00
         9.950205588597E-01    1.000037695846E+00   << s
         9.950205588597E-01    1.000000467791E+00   << s
s >>     9.999999999559E-01    1.000000467791E+00
         9.999999999559E-01    1.000000004956E+00   << *

computed root = 9.999999999559E-01
```

This should be compared with the output from procedure bisect in example 6.1. □

6.4   BATTERY TESTING OF INTERVAL CODES

We can carry out a fairly stringent program of tests and measure the relative performance of the library routines bisect and bisecant on a set of test functions suggested by Bus and Dekker (1975). The functions are

(1) $f(x) = 1 + 2 (xe^{-n} - e^{-nx})$, $n = 1, 5, 10$;
(2) $f(x) = (1 + (1 - n)^2) x - (1 - nx)^2$, $n = 1, 5, 10$;
(3) $f(x) = x^2 - (1 - x)^n$, $n = 1, 5, 10$;
(4) $f(x) = (x - 1) e^{-nx} + x^n$, $n = 1, 5, 10$.

For $n = 5$ and 10 the functions of subset (2) have a turning point and those of subset (3) have a point of inflexion in (0, 1). Subset (4) represents a family of curves increasingly close to the x axis as n increases. In all cases the functions are continuous and change sign over [0, 1]; this is taken as the initial interval.

The test program is written so that a global variable iselect runs over the values 1, 2, 3, 4, and the function parameter n (also defined globally) assumes the values 1, 5, 10. A variable count is used to record the number of function calls and is initialised in the main program for each individual problem. The function f, which is required as a parameter to the library routine, is declared as follows:

```
FUNCTION f ( x : real ) :  real ;
(* The test function selected by the global variable iselect
   is evaluated for the argument x (the function parameter n
   is defined globally). The function evaluation counter count
   is incremented *)
   BEGIN
   count := count + 1 ;
   (* Evaluate one of the following functions *)
   CASE iselect OF
      1:  f := 1 + 2 * (x*exp(-n) - exp(-n*x)) ;
      2:  f := (1 + sqr(1-n)) * x - sqr(1-n*x) ;
      3:  f := sqr(x) - poweri(1-x, n) ;
      4:  f := (x-1)*exp(-n*x) + poweri(x, n)
   END (* CASE *)
END (* f *)
```

When f(x) is invoked, the CASE statement selects the appropriate function, which is then evaluated for n and the supplied argument x.

We tabulate the test results for bisect, bisecant and a further improved interval code called bisrat to be described in section 6.5 (procedure bisrat, based on bisection and rational interpolation, makes use of three successive iterates instead of two as in the secant method). The results for all three library routines to full machine accuracy are

initial interval   0.000000000000E+00   1.000000000000E+00

function	n	function calls			computed root
		bisect	bisecant	bisrat	
1	1	53	8	7	0.4224777096412
	5	55	10	9	0.1382571550568
	10	56	12	10	0.0693140886870
2	1	54	10	9	0.3819660112501
	5	57	11	11	0.0384025518406
	10	59	11	12	0.0099000099980

```
3          1         53         10          9        0.6180339887499
           5         54         10         10        0.3459548158482
          10         55         12         11        0.2451223337533
4          1         54         10          8        0.4010581375415
           5         53         10          9        0.5161535187579
          10         53         10         10        0.5395222269084

total :             656        124        115
```

6.5   RATIONAL INTERPOLATION: PROCEDURE BISRAT

The secant method developed in section 6.2 employs linear interpolation: a non-linear function $f(x)$ is approximated by a straight line passing through two points $(x_k, f(x_k)), (x_{k-1}, f(x_{k-1}))$ on the curve $y = f(x)$. An obvious extension is to use three points instead of two. If we make use of the previous iterate x_{k-2}, more powerful interpolation formulae become available and we expect to obtain methods with a higher asymptotic order of convergence. For example, we can construct the interpolating quadratic for three points and take the zero of the quadratic which lies closer to x_k as the next iterate x_{k+1}. This is the basis of *Muller's method* (see, for example, Conte and de Boor (1980), chapter 3).

A widely used alternative method is given by the following formulae, which can be regarded as a modification of the secant method based on a more accurate estimate of the gradient.

$$
\left.
\begin{aligned}
g_1 &= \frac{f(x_k) - f(x_{k-1})}{x_k - x_{k-1}} \\[2mm]
g_2 &= \frac{f(x_{k-1}) - f(x_{k-2})}{x_{k-1} - x_{k-2}} \\[2mm]
grad &= \frac{f(x_k) - (g_1/g_2) f(x_{k-2})}{x_k - x_{k-2}} \quad \text{provided } g_2 \neq 0 \\[2mm]
x_{k+1} &= x_k - \frac{f(x_k)}{grad} \quad \text{provided } grad \neq 0 \\[2mm]
&(k = 2, 3, \ldots)
\end{aligned}
\right\} \quad (6.2)
$$

These formulae can be derived by interpolating $f(x)$ at x_k, x_{k-1}, x_{k-2} by the *rational function*

$$
g(x) = \frac{x + a}{bx + c}
$$

and taking the zero of $g(x)$ as the next iterate, that is, $x_{k+1} = -a$ (see Dahlquist and Björck (1974), chapter 6). We call this method *3-point rational interpolation*. Three starting points are required, and at each stage a memory of two preceding iterates must be kept.

Rational interpolation can be combined with bisection to produce an interval code with a structure similar to that of bisect and bisecant. The code we give here is based on 'Algorithm R' described by Bus and Dekker (1975). The procedure heading and declarations are

```
PROCEDURE bisrat ( FUNCTION f(x:real) : real ;
                            a, b     : real ;
                            tol      : real ;
                        VAR root     : real ;
                        VAR eflag    : integer ) ;
(* A combination of bisection and 3-point rational interpolation
   is used to compute a root of f(x)=0 in the interval [a, b],
   where f(a)*f(b) <= 0. The routine is based on an algorithm
   described by Bus and Dekker. The parameters and user
   specifications are as for PROCEDURE bisect.
   Routines called from the library:
   sign, errormessage, tolmin           *)

VAR
   x, xmem, xmemold, fx, fmem, fmemold, fa, fb,
   g1, g2, grad, tolx, halfinterval, correction  : real ;
   esteps                                         : integer ;
   converged, limitp, firststep                   : boolean ;
```

The variables xmemold, g1 and g2 are self-explanatory; the variable halfinterval is set to $0.5*(b-a)$ in the iteration loop; esteps records the number of consecutive 'extrapolation steps' as explained below; firststep is initialised to true and switched to false after the first step of the algorithm. The test for change of sign of f and the initialisation of x, xmem etc. are as in bisecant.

Before we develop the body of bisrat, we point out a deficiency in our earlier routine bisecant. If we compute the zero of x^3 for the initial interval $[-1.0, 10.0]$, the following number of function evaluations are required (zero computed to limiting precision):

bisect: 87 bisecant: 206

The reason for the poor performance of bisecant is that, close to a multiple root, the secant method is reduced to first-order convergence. The iteration converges slowly from one side of the root, accumulating a large number of small corrections (extrapolation or *E steps*). To correct for this it would be necessary to interrupt the sequence and, for example, force a bisection step when a certain number of E steps have been performed. A similar behaviour can occur with rational interpolation and should be guarded against in our new code. This is accomplished as follows.

In bisrat it is not assumed that $a \leqslant b$. Instead, the reduction of the interval is performed so that the end-point a always coincides with x, the current estimate of the root. The correction step is then arranged so that the new estimate x lies between a and a + halfinterval. This leads to two possible situations for narrowing the interval (see figure 6.3 — for definiteness we have drawn $a < b$).

In case (i) the reduced interval is $[a, x]$, where $|x - a| \leqslant 0.5 |b - a|$; this is called an intrapolation or *I step* (in the code which follows, b is then set to a and a to x). In case (ii) the reduced interval is $[x, b]$, where $|b - x| > 0.5 |b - a|$; this is called an extrapolation or *E step*. Clearly, an E step will narrow the interval less rapidly than a bisection step; if a number of E steps are performed in

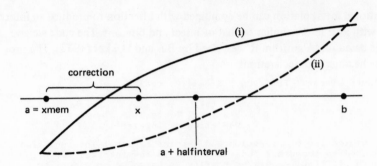

Figure 6.3   Narrowing the interval in procedure bisrat

succession, the sequence should be interrupted and a bisection step forced. An asymptotic analysis (Bus and Dekker (1975)) shows that the appropriate number of E steps to allow is three. If three consecutive E steps have been recorded in the counter esteps, then the correction on the following step is doubled; and if this still leads to an E step, bisection is forced on the next following step.

The worst case is therefore a sequence of the form EEEEB. It follows that if t steps are required to compute a root to a specified accuracy in bisect, then the number of steps required in bisrat is at most $5t$. This is an upper bound; normally, of course, the number is much smaller. On the other hand, the number of steps required in bisecant can be as much as $|b - a|$ /tol (approximately).

The iteration loop for bisrat, incorporating the essential differences from bisecant, is now given.

```
WHILE NOT converged AND NOT limitp DO
    BEGIN  (* iteration loop *)
    halfinterval := 0.5*(b-a) ;
    correction := halfinterval ;  (* bisection as default *)
    IF firststep THEN
        firststep := false
    ELSE  (* bisection forced after 4 consecutive E steps *)
        IF esteps < 4 THEN
        BEGIN  (* rational interpolation *)
        g2 := (fmem/fmemold - 1.0) / (xmem - xmemold) ;
        IF abs(g2) > smallreal THEN
            BEGIN
            g1 := (fx/fmem - 1.0) / (x - xmem) ;
            grad := (fx/fmem - g1/g2) * (fmem/(x-xmemold)) ;
            IF abs(grad) > smallreal THEN
                BEGIN
                correction := -fx/grad ;
                IF esteps = 3 THEN
                    correction := 2*correction ;
                IF abs(correction) < tolx THEN
                    correction := sign(correction)*tolx ;
                (* Use bisection if the new value for x would
                   not lie between x and x+halfinterval *)
                IF (sign(correction) <> sign(halfinterval))
                OR (abs(correction) > abs(halfinterval))  THEN
                    correction := halfinterval
                END
            END
        END  (* rational interpolation *) ;
```

```
(* Update the three iterates *)
xmemold := xmem ;   xmem := x ;
fmemold := fmem ;   fmem := fx ;
x := x + correction ;
(* x is the new estimate of the root *)
fx := f(x) ;

(* Narrow the interval containing the root *)
IF sign(fa)*fx < 0.0 THEN
   BEGIN  (* root lies between a and x *)
   b := a ;    a := x ;
   fb := fa ;  fa := fx ;
   esteps := 0  (* an I step or a bisection step *)
   END
ELSE
   BEGIN  (* root lies between x and b *)
   a := x ;  fa := fx ;
   IF correction = halfinterval THEN
      BEGIN  (* a bisection step *)
      xmemold := xmem ;   xmem := b ;
      fmemold := fmem ;   fmem := fb ;
      esteps := 0
      END
   ELSE
      esteps := esteps + 1  (* an E step *)
   END ;

tolx := tol + tolmin(x) ;
converged := abs(b-a) < tolx ;
limitp := abs(fx) < smallreal
END  (* iteration loop *)
```

The final section of the code, which sets the values of root and eflag, is the same as for bisecant.

Notes

(i)   The formulae used for rational interpolation, involving g1, g2 and grad, are algebraically equivalent to formulae 6.2 but are written in a form which avoids underflow failure.

(ii)   The original Bus and Dekker algorithm has a further refinement in which x (and a) are arranged to lie at the end of the interval where f is smaller in absolute value. Since this appears to give no significant improvement on a large set of test problems, while complicating the code, it has been omitted from bisrat.

Finally, we give comparative results for our three interval codes on a set of functions with zeros of high multiplicity. With initial interval $[-1.0, 10.0]$, the number of function evaluations required to compute a zero to limiting precision is

Function	bisect	bisecant	bisrat
x^3	87	206	158
x^9	32	236	70
x^{19}	17	243	35
x^{25}	12	245	30
Total	148	930	293

EXERCISES

6.1 Modify procedure bisect to output a trace showing the end-points of the interval at each step; hence reproduce the output given in example 6.1. Use the procedure on the same test function with starting intervals $[-1, 5]$ and $[-1, 0.5]$, taking tol = 10^{-2}, 10^{-6} and 0.0, and explain the results.

6.2 Write a procedure for the bisection/Newton method, taking procedure bisect as a model. Use both procedures to solve

(i) $x = e^{-x}$
(ii) $x = \cos x$
(iii) $\ln x = 1/x$

to full machine accuracy, and compare the efficiency of the two routines.

6.3 Devise two formulae based on Steffensen's method and the secant method to compute the nth root of a real number $c > 0$. Hence, using a calculator, determine $e^{1/3}$ and $\pi^{1/5}$ correct to 4 decimal places. Compare with the performance of Newton's method in exercise 5.1.

6.4 Modify procedure bisecant to output a trace showing the end-points of the interval and indicating the type of step used; hence reproduce the output given in example 6.3. Use the procedure to solve the equations in exercise 6.2, and compare with the performance of the two earlier routines.

6.5 Determine all the roots of the equation

$$x/\sin x = e^{x/\tan x}$$

between 0 and 20 to an accuracy of 6 decimal places. Can you formulate a conjecture on the approximate location of the root which lies closest to 50? Test your conjecture, and determine the root.

6.6 Near a multiple root the secant method converges from one side of the root, accumulating a large number of small corrections, and convergence is very slow. Modify procedure bisecant to test whether x and xmem lie on the same side of the root; if two consecutive secant steps have this property, force a bisection step in place of the second secant step. Compare the old and new verions of bisecant on the test functions x^5, x^9 and x^{13} with initial interval $[-1, 10]$.

6.7 Write a battery-testing program for an interval code on the lines described in section 6.4. Test the procedures bisect and bisecant on the four groups of functions given there (the output need not be set out in the same form as in the table). Are your results in agreement with the table? Explain any discrepancies.

6.8   Run the battery test of exercise 6.7 on procedure bisrat. Use bisrat to solve the equations in exercise 6.2, and compare with the performance of the earlier routines.

7 Systems of Linear Equations

Our little systems have their day.

Lord Tennyson, *In Memoriam*

One of the commonest problems of numerical computation is to solve a *system of simultaneous linear equations*

$$\left.\begin{array}{c} a_{11}x_1 + a_{12}x_2 + \ldots + a_{1n}x_n = b_1 \\ a_{21}x_1 + a_{22}x_2 + \ldots + a_{2n}x_n = b_2 \\ \cdot \quad \cdot \quad \cdot \quad \cdot \quad \cdot \quad \cdot \quad \cdot \\ \cdot \quad \cdot \quad \cdot \quad \cdot \quad \cdot \quad \cdot \quad \cdot \\ a_{n1}x_1 + a_{n2}x_2 + \ldots + a_{nn}x_n = b_n \end{array}\right\} \tag{7.1}$$

The coefficients a_{ij} for $1 \leqslant i, j \leqslant n$ and the right hand sides b_i for $1 \leqslant i \leqslant n$ are given; the problem is to find numerical values for the unknowns $x_1, \ldots, x_n$ which satisfy the n equations.

Systems of linear equations often arise as a result of linearising approximations made in harder problems, such as fitting curves to a set of data points or solving boundary value problems in differential equations. It is therefore important that we have accurate and efficient library routines to solve this basic problem. The method of solution depends on the coefficient matrix. If the order n is not too large, *direct elimination methods* are used. In some important applications, however, linear systems arise with n running into thousands and most of the coefficients zero; such *sparse* systems are often solved using *iterative methods* to avoid storing and manipulating very large matrices.

The library routines to be developed in this chapter fall into the above two classes. The procedures lufac, lusolv and linsolv in sections 7.2 and 7.3 are based on *Gaussian elimination*, a direct method. The procedure congrad in section 7.6 is based on the *conjugate gradient method*, in effect an iterative method for solving large, sparse systems.

7.1 BASIC CONCEPTS

We begin by reviewing the notation and some definitions of matrix theory, which underlies the study of linear algebra. A matrix is a rectangular array of elements;

we shall take the elements to represent real numbers and use the following notation.

m × *n matrix* $A = (a_{ij})$, $i = 1, \ldots, m$, $j = 1, \ldots, n$.

An *m* × *n* matrix has *m rows* and *n columns*. The element a_{ij} at the intersection of the *i*th row and *j*th column is referred to as the (i, j)th element. If $m = n$, the matrix is said to be *square* of order *n*.

Diagonal matrix $D = (d_{ij})$: a square matrix such that $d_{ij} = 0$ for $i \neq j$ (the off-diagonal elements are all zero). The *unit matrix* of order *n* is the *n* × *n* diagonal matrix $I = (e_{ij})$ with $e_{ii} = 1$ for $1 \leqslant i \leqslant n$.

Lower triangular matrix $L = (l_{ij})$: a square matrix such that $l_{ij} = 0$ for $i < j$.

Upper triangular matrix $U = (u_{ij})$: a square matrix such that $u_{ij} = 0$ for $i > j$.

These definitions are illustrated in the following examples.

$$D = \begin{bmatrix} 2 & 0 & 0 \\ 0 & 0 & 0 \\ 0 & 0 & -1 \end{bmatrix}, \; L = \begin{bmatrix} 1 & 0 & 0 \\ -4 & 1 & 0 \\ 2 & -1 & 1 \end{bmatrix}, \; U = \begin{bmatrix} 2 & 3 & 1 \\ 0 & -1 & 2 \\ 0 & 0 & 5 \end{bmatrix}$$

(here L is called *unit* lower triangular since $l_{ii} = 1$ for $1 \leqslant i \leqslant n$).

Matrix multiplication The *product* of an *m* × *p* matrix A and a *q* × *n* matrix B is defined if and only if $p = q$; then $C = AB$, where $C = (c_{ij})$ is given by

$$c_{ij} = \sum_{k=1}^{p} a_{ik} b_{kj} \text{ for } i = 1, \ldots, m, \; j = 1, \ldots, n$$

The product C is an *m* × *n* matrix. In general, $BA \neq AB$ even if the products are defined. However, we have $IA = AI = A$, where A is a square matrix and I is the unit matrix of the same order. Matrix multiplication is associative, that is, $(AB)C = A(BC)$.

Transpose $A^T = (a_{ji})$: the matrix obtained from A by writing the rows as columns or vice versa. From the matrix multiplication rule it can be shown that if the product $A_1 A_2 \ldots A_r$ is defined, then $(A_1 A_2 \ldots A_r)^T = A_r^T \ldots A_2^T A_1^T$. A square matrix A is said to be *symmetric* if $A^T = A$.

Inverse An inverse of an *n* × *n* matrix A is an *n* × *n* matrix A^{-1}, if one exists, such that $AA^{-1} = I$. It can be shown that an inverse of A, if it exists, is unique and also satisfies $A^{-1}A = I$. It follows from the definition that if $A_1, A_2, \ldots, A_r$ are *n* × *n* matrices each with an inverse, then $(A_1 A_2 \ldots A_r)^{-1} = A_r^{-1} \ldots A_2^{-1} A_1^{-1}$.

Column vector   A column vector is an $n \times 1$ matrix

$$\mathbf{v} = \begin{bmatrix} v_1 \\ \cdot \\ v_i \\ \cdot \\ v_n \end{bmatrix}$$

often written as the transpose of a *row vector*, $\mathbf{v} = (v_1, \ldots, v_i, \ldots, v_n)^T$. The element v_i is called the *ith component* of $\mathbf{v}$. If all the components are zero, the vector is called the *zero* or *null* vector, $\mathbf{0} = (0, \ldots, 0)^T$.

Inner product   If $\mathbf{u}$ and $\mathbf{v}$ are two n-component column vectors, then the matrix product $\mathbf{u}^T\mathbf{v}$ is defined and is given by

$$\mathbf{u}^T\mathbf{v} = \sum_{i=1}^{n} u_i v_i$$

The product $\mathbf{u}^T\mathbf{v}$ is a 1×1 matrix (a *scalar* or single number); it is called the *inner product* of $\mathbf{u}$ and $\mathbf{v}$.

If $\mathbf{A}$ is an $m \times n$ matrix and $\mathbf{x}$ is an n-component column vector, then the matrix product $\mathbf{Ax}$ is defined and is given by

$$(\mathbf{Ax})_i = \sum_{j=1}^{n} a_{ij}x_j \quad \text{for} \quad i = 1, \ldots, m$$

The product $\mathbf{Ax}$ is an m-component column vector. For example

$$\begin{bmatrix} 1 & 2 & -1 \\ 3 & -1 & 4 \end{bmatrix} \begin{bmatrix} x_1 \\ x_2 \\ x_3 \end{bmatrix} = \begin{bmatrix} x_1 + 2x_2 - x_3 \\ 3x_1 - x_2 + 4x_3 \end{bmatrix} \quad (m = 2, n = 3)$$

In matrix notation the system of linear equations 7.1 may be written as

$$\mathbf{Ax} = \mathbf{b} \tag{7.2}$$

where $\mathbf{A} = (a_{ij})$ is the $n \times n$ matrix of coefficients, $\mathbf{b}$ is the n-component column vector of right hand sides and $\mathbf{x}$ is an n-component column vector of unknowns. Since $\mathbf{Ax}$ and $\mathbf{b}$ are column vectors, equation 7.2 represents n simultaneous equations as in the system of equations 7.1 (the reader is recommended to verify this). Our problem is to solve the system for $\mathbf{x}$. We assume, to begin with, that a solution exists.

Solving $\mathbf{Ax} = \mathbf{b}$ is easy if the matrix is upper or lower triangular. If $\mathbf{A} = \mathbf{U}$, an upper triangular matrix, then

$$\left. \begin{array}{l} u_{11}x_1 + u_{12}x_2 + \ldots + u_{1n}x_n = b_1 \\ \qquad u_{22}x_2 + \ldots + u_{2n}x_n = b_2 \\ \qquad\qquad \cdot \qquad \cdot \qquad \cdot \\ \qquad\qquad\quad \cdot \qquad \cdot \qquad \cdot \\ \qquad\qquad\qquad\quad \cdot \qquad \cdot \\ \qquad\qquad\qquad\qquad u_{nn}x_n = b_n \end{array} \right\} \tag{7.3}$$

Provided $u_{ii} \neq 0$ for $1 \leqslant i \leqslant n$, the system of equations 7.3 can be solved by *back substitution*. Starting on the bottom row and working upwards, we determine the components of **x** in reverse order

$$x_n = b_n/u_{nn}$$

$$x_i = (b_i - \sum_{j=i+1}^{n} u_{ij}x_j)/u_{ii}, \quad i = n-1, \ldots, 1 \qquad (7.4)$$

At each stage of this process the components of **x** required on the right hand side of equation 7.4 are available, having been determined at an earlier stage. A similar method using *forward substitution* can be employed if the matrix is lower triangular (again it is necessary that all the diagonal elements be non-zero).

We now show how to transform a full system of the form 7.1 into an equivalent triangular system of the form 7.3 (by 'equivalent' we mean that the two systems have the same solution, which we aim to find).

Gaussian Elimination

Gaussian elimination is a method of reducing an $n \times n$ system of linear equations to upper triangular form. If the coefficient a_{11} in the system of equations 7.1 is non-zero, we subtract a multiple a_{i1}/a_{11} of the first equation from the ith equation for $i = 2, \ldots, n$, being careful to perform the same operations on the right hand side. The effect is to eliminate x_1 from all equations below the first. In matrix terminology we obtain a coefficient matrix with zeros in the first column below a_{11}, all elements from the second row downwards modified, and a modified right hand side. The element a_{11} is called the *pivot*; the ratios $m_{i1} = a_{i1}/a_{11}$ for $i = 2, \ldots, n$ are called the *multipliers* for the respective rows.

We next take as pivot the element in the $(2, 2)$ diagonal position of the modified matrix, supposing this to be non-zero, and eliminate x_2 from equations $3, \ldots, n$. In general, pivoting on the (k, k)th diagonal element, we eliminate x_k from equations $k + 1, \ldots, n$. We continue until $k = n - 1$, when x_{n-1} is finally eliminated from the nth equation and we have an upper triangular system.

Example 7.1

Reduce the following system to upper triangular form by Gaussian elimination.

$$\begin{bmatrix} 2 & -2 & 4 \\ 1 & 4 & 3 \\ 5 & 0 & 2 \end{bmatrix} \begin{bmatrix} x_1 \\ x_2 \\ x_3 \end{bmatrix} = \begin{bmatrix} 4 \\ 3 \\ 10 \end{bmatrix}$$

$k = 1$: with $m_{21} = 1/2$ and $m_{31} = 5/2$ we obtain

$$\begin{bmatrix} 2 & -2 & 4 \\ 0 & 5 & 1 \\ 0 & 5 & -8 \end{bmatrix} \begin{bmatrix} x_1 \\ x_2 \\ x_3 \end{bmatrix} = \begin{bmatrix} 4 \\ 1 \\ 0 \end{bmatrix}$$

$k = 2$: with $m_{32} = 5/5 = 1$ we obtain the triangular system

$$\begin{bmatrix} 2 & -2 & 4 \\ 0 & 5 & 1 \\ 0 & 0 & -9 \end{bmatrix} \begin{bmatrix} x_1 \\ x_2 \\ x_3 \end{bmatrix} = \begin{bmatrix} 4 \\ 1 \\ -1 \end{bmatrix}$$

In this form the system could be solved by back substitution; the unknowns would be determined in the order x_3, x_2, x_1. □

In figure 7.1 we show the situation at the kth stage of Gaussian elimination when the pivot is the element in the (k, k)th diagonal position. The superscript $(k - 1)$ indicates that the elements shown have been modified by $k - 1$ previous elimination steps.

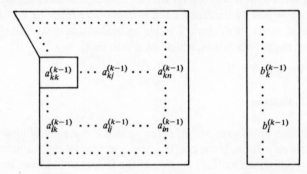

Figure 7.1 kth stage of Gaussian elimination

The operations for the kth elimination stage are

FOR $i := k + 1$ TO n DO

    BEGIN

    $m_{ik} := a_{ik}^{(k-1)}/a_{kk}^{(k-1)}$;   (* multiplier for ith row *)

    FOR $j := k + 1$ TO n DO

        $a_{ij}^{(k)} := a_{ij}^{(k-1)} - m_{ik} a_{kj}^{(k-1)}$;   (* modify ith row *)

    $b_i^{(k)} := b_i^{(k-1)} - m_{ik} b_k^{(k-1)}$;   (* modify RHS *)

    END   (7.5)

(the zeros in the kth column are not computed explicitly).

This might be coded in Pascal as follows.

```
pivot := A[k,k] ;
FOR i := k+1 TO n DO
   BEGIN
   multiplier := A[i,k]/pivot ;
   FOR j := k+1 TO n DO
      A[i,j] := A[i,j] - multiplier*A[k,j] ;
   b[i] := b[i] - multiplier*b[k]
   END
```

We shall develop a number of refinements before giving the final version of this code in section 7.2.

Gaussian elimination is a *direct method* which terminates after a fixed number of steps, so it is possible to calculate the number of arithmetic operations required for the whole process. To analyse the efficiency of the algorithm we need to know how the number of operations, and hence the expected computation time, depends on n, the order of the system. From operations 7.5 we see that at the kth elimination stage the operations concerned with row i $(i > k)$ are: 1 division, $n - k + 1$ multiplications, $n - k + 1$ subtractions. There are $n - k$ rows below the pivot, hence the total number of multiplications (and subtractions) at the kth elimination stage is $(n - k)(n - k + 1)$.

The number of multiplications (and subtractions) for the whole process is therefore

$$\sum_{k=1}^{n-1} (n - k)(n - k + 1)$$

$$= \sum_{k=1}^{n-1} k^2 + n(n - 1)/2$$

$$= \tfrac{1}{3}n^3 + 0(n^2) \tag{7.6}$$

To complete the solution we use the back substitution process from equation 7.4, which requires $1 + 2 + \ldots + (n - 1) = n(n - 1)/2$ multiplications, the same number of subtractions or additions, and n divisions.

The important conclusion is that most of the work in solving a linear system by elimination is performed in the actual elimination process. The time required for elimination increases asymptotically as n^3. By comparison, the time required for the substitution process, which is $0(n^2)$, is negligible. For example, to solve a general 100×100 system would require a little over 1/3 million 'multiplicative' operations and a similar number of 'additive' operations; over 98 per cent of these operations are accounted for in the elimination process.

Singular Systems

The inverse of an $n \times n$ matrix $\mathbf{A}$ has been defined as an $n \times n$ matrix $\mathbf{A}^{-1}$ satisfying $\mathbf{A}\mathbf{A}^{-1} = \mathbf{A}^{-1}\mathbf{A} = \mathbf{I}$. However, not all square matrices possess inverses. A familiar example is the 1×1 matrix $[0]$, for which there is certainly no 1×1 matrix $[a]$ such that $[0]$ $[a] = [1]$. A slightly less trivial case is

$$\begin{bmatrix} 4 & 6 \\ 2 & 3 \end{bmatrix}$$

which also does not possess an inverse (we shall return to this in example 7.2). Such matrices are said to be *singular*. If, on the other hand, the inverse does exist, then the matrix is said to be *non-singular*.

The relevance of singularity to our problem of solving linear systems is contained in the following theorem.

Theorem 7.1

An $n \times n$ system $\mathbf{Ax} = \mathbf{b}$ has a solution $\mathbf{x}$ for arbitrary right hand side $\mathbf{b}$ if and only if $\mathbf{A}$ is non-singular (the solution is then unique).

Proof

(i)   If $\mathbf{A}$ is non-singular then $\mathbf{A}^{-1}$ exists. Premultiply $\mathbf{Ax} = \mathbf{b}$ by $\mathbf{A}^{-1}$ to obtain

$$\mathbf{A}^{-1}\mathbf{Ax} = \mathbf{A}^{-1}\mathbf{b}$$

that is

$$\mathbf{x} = \mathbf{A}^{-1}\mathbf{b}$$

We see that $\mathbf{A}^{-1}$ may be regarded as an operator which transforms any given right hand side $\mathbf{b}$ into the corresponding solution $\mathbf{x}$. Since $\mathbf{A}^{-1}$ exists and is unique, the same is true of $\mathbf{x}$.

(ii)   If the system $\mathbf{Ax} = \mathbf{b}$ has a solution for arbitrary $\mathbf{b}$, then in particular it has a solution when $\mathbf{b}$ is the unit vector $\mathbf{e}_i$ (defined as the column vector with ith component equal to 1, the other components zero). Thus $\mathbf{Ax}_i = \mathbf{e}_i$ for some column vector $\mathbf{x}_i$. Taking $i = 1, \ldots, n$, and setting the columns $\mathbf{x}_i$ side by side and similarly the columns $\mathbf{e}_i$, we can claim $\mathbf{AX} = \mathbf{I}$ for some $n \times n$ matrix $\mathbf{X}$. However, $\mathbf{X}$ is then the inverse of $\mathbf{A}$ by definition, hence $\mathbf{A}$ is non-singular. □

Theorem 7.1 is an existence and uniqueness theorem for solutions of an $n \times n$ system $\mathbf{Ax} = \mathbf{b}$. We have a method of solving such a system by Gaussian elimination and back substitution. This prompts us to consider how the process of solving $\mathbf{Ax} = \mathbf{b}$ may break down. There is just one way in which this can happen. If a pivot $a_{kk}^{(k-1)}$ in operations 7.5 is zero, then the multipliers $m_{ik} = a_{ik}^{(k-1)}/a_{kk}^{(k-1)}$ cannot be formed and elimination in the kth column cannot proceed. If there is a lower row (say the lth) with a non-zero element in the (l, k) position, then we can interchange rows k and l and take this non-zero element as pivot (if the components on the right hand side are interchanged as well, the solution is unaffected). If, however, $a_{kk}^{(k-1)}$ is zero and all the elements below it in the kth column are also zero, then no possible pivot can be found. We may consider interchanging columns, which corresponds to reordering the unknowns, but this would only postpone the problem to a later stage.

From the point of view of elimination it does not matter if the pivotal sub-column is all zero, since for that very reason no eliminations are needed in the kth column. We could continue and still obtain an upper triangular system, but the element in the (k, k)th position of the matrix $\mathbf{U}$ would be zero; back substitution would then certainly fail on division by zero. We summarise this as an important corollary of theorem 7.1.

Corollary 7.1'

An $n \times n$ matrix **A** is singular if and only if the upper triangular matrix **U** produced by Gaussian elimination with row interchanges has at least one zero element on the diagonal. □

In library routines we aim only to solve non-singular systems, each of which, by theorem 7.1, has a unique solution. Corollary 7.1' will be used to detect singularity; this will be treated as an error condition in our routines. We remark in passing that if the coefficient matrix is singular then there are two possible cases for the solution, depending on the right hand side. We show these by an example.

Example 7.2

Solve the system†

$$4x + 6y = 4$$
$$2x + 3y = 2 + \epsilon$$

where ϵ is a constant.

We obtain by elimination

$$4x + 6y = 4$$
$$0 = \epsilon$$

Case (i) $\epsilon \neq 0$: no solution exists which will satisfy both equations; the system is said to be *inconsistent.*

Case (ii) $\epsilon = 0$: the equations are consistent, in fact one is a multiple of the other. The best we can do is solve for one unknown in terms of the other, for example, $y = 2(1 - x)/3$; there are an infinite number of solutions dependent on an undetermined parameter x. □

7.2   TRIANGULAR FACTORISATION

We recall from equation 7.6 that most of the work in solving a linear system by elimination and substitution is done in the elimination process and much less in the substitution process. In practice we often have to solve systems with the same coefficient matrix and several or many different right hand sides. Is there any way to save the results of elimination and use these repeatedly? The multipliers m_{ij} and the upper triangular matrix **U** produced by elimination depend only on the coefficient matrix. If we save the multipliers and **U**, we can solve a system with the same coefficient matrix and any right hand side without having to rework the elimination.

†The coefficient matrix in the example is singular. The rows (and columns) are *linearly dependent* (in the 2×2 case this means that one row is a multiple of the other); also the *determinant* is $4 \times 3 - 6 \times 2 = 0$. These are two further conditions equivalent to singularity.

Example 7.3

Solve the system $\mathbf{Ax} = \mathbf{b}$, where

$$\mathbf{A} = \begin{bmatrix} 2 & -2 & 4 \\ 1 & 4 & 3 \\ 5 & 0 & 2 \end{bmatrix}, \ \mathbf{b} = \begin{bmatrix} 8 \\ 0 \\ 7 \end{bmatrix}$$

Gaussian elimination for the matrix $\mathbf{A}$ has been carried out already in example 7.1. The multipliers were found to be $m_{21} = 1/2, m_{31} = 5/2, m_{32} = 1$. We arrange these in a unit lower triangular matrix and also recall the upper triangular matrix

$$\mathbf{L} = \begin{bmatrix} 1 & 0 & 0 \\ 1/2 & 1 & 0 \\ 5/2 & 1 & 1 \end{bmatrix}, \ \mathbf{U} = \begin{bmatrix} 2 & -2 & 4 \\ 0 & 5 & 1 \\ 0 & 0 & -9 \end{bmatrix}$$

The right hand side $\mathbf{b}$ is now modified according to the steps performed in the elimination process, as recorded by the multipliers m_{ij}.

$$\begin{bmatrix} 8 \\ 0 \\ 7 \end{bmatrix} \rightarrow \begin{bmatrix} 8 \\ 0 - 1/2 \times 8 = -4 \\ 7 - 5/2 \times 8 = -13 \end{bmatrix} \rightarrow \begin{bmatrix} 8 \\ -4 \\ -13 - 1 \times (-4) = -9 \end{bmatrix}$$

The reduced triangular system is

$$\begin{bmatrix} 2 & -2 & 4 \\ 0 & 5 & 1 \\ 0 & 0 & -9 \end{bmatrix} \begin{bmatrix} x_1 \\ x_2 \\ x_3 \end{bmatrix} = \begin{bmatrix} 8 \\ -4 \\ -9 \end{bmatrix}$$

and this can be solved by back substitution to give

$x_3 = 1$
$x_2 = (-4 - 1 \times 1)/5 = -1$
$x_1 = (8 + 2 \times (-1) - 4 \times 1)/2 = 1$ □

The first stage of Gaussian elimination consists in subtracting a multiple $m_{i1} = a_{i1}/a_{11}$ of the first equation from the ith equation for $i = 2, \ldots, n$. The right hand side is modified to give

$$b_i^{(1)} = b_i - m_{i1}b_1 \quad \text{for} \quad i = 2, \ldots, n$$

This can be expressed in a useful form as a matrix product

$$\begin{bmatrix} b_1 \\ b_2 - m_{21}b_1 \\ \cdot \\ \cdot \\ \cdot \\ b_n - m_{n1}b_1 \end{bmatrix} = \begin{bmatrix} 1 & 0 & \cdot & \cdot & \cdot & 0 \\ -m_{21} & 1 & 0 & \cdot & \cdot & 0 \\ \cdot & & 0 & \cdot & & \cdot \\ \cdot & & & \cdot & & \cdot \\ \cdot & & & & \cdot & \cdot \\ -m_{n1} & 0 & \cdot & \cdot & \cdot & 1 \end{bmatrix} \begin{bmatrix} b_1 \\ b_2 \\ \cdot \\ \cdot \\ \cdot \\ b_n \end{bmatrix}$$

The effect of the first elimination stage on the right hand side is expressed by the transformation $b^{(1)} = M_1 b$, where M_1 is the matrix above. Similarly, the effect of the kth elimination stage can be expressed as $b^{(k)} = M_k b^{(k-1)}$, where

$$
M_k = \begin{bmatrix}
1 & & & & & & & & & & \\
0 & & \cdot & \cdot & & & & & & & \\
0 & & & \cdot & 1 & & & & & & \\
\cdot & & & & 0 & 1 & & & & & \\
\cdot & & & & 0 & -m_{k+1,k} & 1 & & & & \\
\cdot & & & & & \cdot & 0 & \cdot & & & \\
\cdot & & & & & \cdot & & \cdot & & & \\
\cdot & & & & & \cdot & & & \cdot & & \\
\cdot & & & & & \cdot & & & & 1 & \\
0 & \cdot & \cdot & \cdot & 0 & -m_{nk} & 0 & \cdot & \cdot & 0 & 1
\end{bmatrix}
$$

(the reader is advised to verify this by reference to the operations in 7.5.

At the end of the elimination process the right hand side is finally the product

$M_{n-1} \ldots M_1 \, b$

and the reduced triangular system therefore takes the form

$U x = M_{n-1} \ldots M_1 \, b$

We can solve the system for any given right hand side b by first forming the modified column vector

$y = M_{n-1} \ldots M_1 \, b$ (7.7)

and then solving the upper triangular system

$U x = y$

In practice y is usually found by solving a system derived from equation 7.7,

$(M_{n-1} \ldots M_1)^{-1} \, y = b$

that is

$M_1^{-1} \ldots M_{n-1}^{-1} \, y = b$

The reason is the following. It is easily verified that

$$
M_k^{-1} = \begin{bmatrix}
1 & & \cdot & & & & & & & & \\
0 & & & \cdot & & & & & & & \\
0 & & & & 1 & & & & & & \\
\cdot & & & & 0 & 1 & & & & & \\
\cdot & & & & 0 & m_{k+1,k} & 1 & \cdot & & & \\
\cdot & & & & & \cdot & & 0 & \cdot & & \\
\cdot & & & & & \cdot & & & \cdot & & \\
\cdot & & & & & \cdot & & & & 1 & \\
0 & \cdot & \cdot & \cdot & 0 & m_{nk} & 0 & \cdot & \cdot & 0 & 1
\end{bmatrix}
$$

and the product $M_1^{-1} \ldots M_{n-1}^{-1}$ is found to be a unit lower triangular matrix

$$L = \begin{bmatrix} 1 & & & & & & \\ m_{21} & 1 & & & & & \\ m_{31} & m_{32} & 1 & \cdot & & & \\ \cdot & \cdot & \cdot & \cdot & \cdot & & \\ \cdot & \cdot & \cdot & & \cdot & & \\ \cdot & \cdot & \cdot & & & 1 & \\ m_{n1} & m_{n2} & \cdot & \cdot & \cdot & m_{n,n-1} & 1 \end{bmatrix}$$

Thus the system $Ly = b$ can be solved easily for y by forward substitution.

We note that since $Ly = b$ and $Ux = y$, it follows that $LUx = b$. In summary, Gaussian elimination *factorises* the matrix A in the form

$A = LU$

where L is a unit lower triangular matrix and U is an upper triangular matrix. The system $LUx = b$ can then be solved in two substitution processes

$Ly = b$ (forward substitution)
$Ux = y$ (back substitution)   (7.8)

If row interchanges are performed in the elimination, then L and U are triangular factors of A', a row-permuted form of A. Provided that b is permuted in the same way, the system can still be solved correctly.

The matrices L and U in example 7.3 are the triangular factors of A produced by Gaussian elimination. It can be checked that $LU = A$!

Implementation of Gaussian Elimination

We now consider in detail how to program Gaussian elimination. We have seen that if a zero pivot is encountered in the elimination process, then it is necessary to interchange rows (if all the potential pivots in a column are zero the matrix is singular). In fact, a pivot with the exact value zero is unlikely to arise in computations with real numbers because of rounding error. We may expect, however, that the occurrence of a pivot which is almost zero will signal difficulties in finite-precision arithmetic.

Example 7.4

Solve the following system in 4-digit floating-point arithmetic.

$$\begin{bmatrix} 1.000 \times 10^{-4} & 1.000 \\ 1.000 & 2.000 \end{bmatrix} \begin{bmatrix} x_1 \\ x_2 \end{bmatrix} = \begin{bmatrix} 1.000 \\ 0.000 \end{bmatrix}$$

The triangular system obtained without row interchange is

$$\begin{bmatrix} 1.000 \times 10^{-4} & 1.000 \\ 0 & -10,000 \end{bmatrix} \begin{bmatrix} x_1 \\ x_2 \end{bmatrix} = \begin{bmatrix} 1.000 \\ -10,000 \end{bmatrix}$$

and the computed solution is $x_2 = 1.000, x_1 = 0.000$.

If we first interchange rows because of the small pivot 1.000×10^{-4}, we obtain after elimination

$$\begin{bmatrix} 1.000 & 2.000 \\ 0 & 1.000 \end{bmatrix} \begin{bmatrix} x_1 \\ x_2 \end{bmatrix} = \begin{bmatrix} 0.000 \\ 1.000 \end{bmatrix}$$

and the computed solution is $x_2 = 1.000, x_1 = -2.000$.

The true solution is $x_2 = 1.0002, x_1 = -2.0004$ correct to 4 decimal places. □

In the first part of example 7.4 the large absolute value of the modified matrix element $a_{22}^{(1)}$ is indicative of serious difficulties in 4-digit arithmetic; the solution obtained is useless. In the second part this catastrophic instability is avoided by interchanging equations, taking 1.000 as pivot and a multiplier $m_{21} = 10^{-4}$; the resulting solution is correct to 4-figure accuracy.

A detailed error analysis (see, for example, Forsythe and Moler (1967), section 21) shows that this is generally true, though usually less dramatic. We must prevent a large growth of the modified matrix elements. An effective way to do this is to choose as pivot the largest element in the pivotal subcolumn. At the kth elimination stage we examine the elements from $a_{kk}^{(k-1)}$ downwards (see figure 7.1) and find the element of largest absolute value; this is then brought up to the pivotal position by an interchange of equations. The multipliers calculated from this pivot necessarily satisfy $|m_{ik}| \leqslant 1$ for $i = k + 1, \ldots, n$. This is the strategy of *partial pivoting*†.

In a computer program it will not be necessary to interchange rows explicitly. We keep a record of the row indices in a vector row, initially in the order $1, 2, \ldots, n$, and instead of interchanging rows of the array we interchange the corresponding components of row. We then refer to the element in the (i, j)th position as A[row[i], j] instead of A[i, j].

Some form of *scaling* should be performed on the original matrix if the coefficients vary considerably in size. We can see that an ill-advised scaling of rows can make nonsense of the partial pivoting strategy by considering again example 7.4. If the first equation is multiplied by 10^5 we have

$$\begin{bmatrix} 10.00 & 100{,}000 \\ 1.000 & 2.000 \end{bmatrix} \begin{bmatrix} x_1 \\ x_2 \end{bmatrix} = \begin{bmatrix} 100{,}000 \\ 0.000 \end{bmatrix}$$

The pivoting strategy now requires no row interchange. We obtain by elimination

$$\begin{bmatrix} 10.00 & 100{,}000 \\ 0 & -10{,}000 \end{bmatrix} \begin{bmatrix} x_1 \\ x_2 \end{bmatrix} = \begin{bmatrix} 100{,}000 \\ -10{,}000 \end{bmatrix}$$

leading to the solution $x_2 = 1.000, x_1 = 0.000$. This is the bad solution of example 7.4.

It is clear that we can force almost any selection of pivots by a suitable pre-scaling of the rows of a matrix, and we must decide on some standardisation to justify the partial pivoting strategy. This is far from easy theoretically. We shall

†Complete pivoting, involving column as well as row interchanges, is rarely called for and will not be considered.

adopt the common compromise of *row equilibration*: each row of the matrix **A** is scaled so that $\max_j |a_{ij}| = 1$. The scaled form of the above matrix is then

$$\begin{bmatrix} 1.000 \times 10^{-4} & 1.000 \\ 0.500 & 1.000 \end{bmatrix}$$

and partial pivoting works as intended.

In a computer program we shall not scale the equations explicitly. Instead, we store the scale factors in a vector scale and use these factors only when searching for pivots. This avoids unnecessary rounding errors which would be introduced by multiplying the matrix elements, and also leaves the right hand side unaffected.

We now give a library routine to compute the triangular factors **L** and **U** of a matrix **A** by Gaussian elimination with row equilibration and partial pivoting. This routine is similar, with some modifications, to an ALGOL procedure given by Forsythe and Moler (1967). The routine does most of the work towards solving a system of linear equations, but the final forward and backward substitution to compute the solution for a given right hand side is left to a later routine.

```
PROCEDURE lufac ( n   : range ;     (* order of the matrix        *)
                  VAR A  : matrix ;   (* input- nxn matrix          *)
                  VAR LU : matrix ;   (* output- triangular factors *)
                  VAR row: ivector ;  (* permuted row indices       *)
                  VAR eflag : integer ) ;

(* This procedure factorises a nonsingular nxn matrix A (with
   row permutation) in the form A' = L*U, where L is a unit
   lower triangular matrix and U is an upper triangular matrix.
   Gaussian elimination with scaling and partial pivoting is
   used. The triangular factors, minus the unit diagonal of L,
   are returned in the array LU. Tne permuted row indices are
   returned in the vector row.
   The error number for eflag is:
     =1  The matrix is singular to within rounding error (fatal).
   Routine called from the library:  errormessage *)

VAR
   max, size, pivot, multiplier : real ;
   i, j, k, l, pivotrow, rowi   : range ;
   singular                     : boolean ;
   scale                        : rvector ;

BEGIN
singular := false ;  (* switched to true if A is
                        found to be singular *)
FOR i := 1 TO n DO
   BEGIN    (* Initialise row index, copy A into LU,
               and calculate row scale factor *)
   row[i] := i ;
   max := 0.0 ;
   FOR j := 1 TO n DO
      BEGIN
      LU[i,j] := A[i,j] ;
      IF max < abs(LU[i,j]) THEN
         max := abs(LU[i,j])
      END ;
   IF max = 0.0 THEN
      singular := true  (* zero row in matrix *)
   ELSE
      scale[i] := 1/max
   END (* i *) ;
```

```
(* Gaussian elimination with partial pivoting *)
k := 1 ;
WHILE NOT singular AND (k < n) DO
    BEGIN    (* seek a pivot in kth column *)
    max := 0.0 ;
    FOR i := k TO n DO
        BEGIN
        size := scale[row[i]] * abs(LU[row[i],k]) ;
        IF max < size THEN
            BEGIN
            max := size ;  l := i
            END
        END ;
    IF max < k*rprec4 THEN
        singular := true  (* pivotal subcolumn all zero
                              to within rounding error *)
    ELSE
        BEGIN    (* a pivot has been found *)
        pivotrow := row[l] ;
        pivot := LU[pivotrow,k] ;
        IF l <> k THEN
            BEGIN    (* interchange indices of rows l and k *)
            row[l] := row[k] ;
            row[k] := pivotrow
            END ;
        (* Elimination using (k,k) pivot *)
        FOR i := k+1 TO n DO
            BEGIN
            rowi := row[i] ;
            multiplier := LU[rowi,k] / pivot ;
            LU[rowi,k] := multiplier ;  (* save multiplier *)
            IF multiplier <> 0.0 THEN
                FOR j := k+1 TO n DO
                    LU[rowi,j] := LU[rowi,j]
                                        - multiplier*LU[pivotrow,j]
            END
        END ;
    k := k + 1
    END (* of kth elimination stage *) ;

IF singular
OR (scale[row[n]]*abs(LU[row[n],n]) < n*rprec4) THEN
    errormessage ('lufac    ', 1,
        'matrix is singular            ', fatal, eflag)
ELSE
    eflag := 0

END (* lufac *)
```

Notes

(i)   Calling lufac (n, A, A, row, eflag) will overwrite A with LU. In some applications (see section 7.3) it is necessary to retain A unmodified; this is why a separate formal parameter has been provided for the array LU.

(ii)   The parameter row returns the permuted row indices; these are required subsequently when a system **LUx = b** is solved.

(iii) Procedure lufac abandons the factorisation and returns with an error message if the matrix is found to be singular or nearly singular. This means either that the matrix contains a zero row or, more likely, that a pivot within the round-off error level has been encountered. Pivots are unlikely to be exactly zero; the test at the kth elimination stage is essentially

$$\max_{k \leqslant i \leqslant n} \; (\text{scale } (i) \times |a_{ik}^{(k-1)}|) < k \times \text{rprec4}$$

The threshold used here takes account of the possible error growth in the elimination process, which for practical purposes can be taken as proportional to k and the relative precision (see Forsythe and Moler (1967), section 21).

(iv)   The error message should read in full: 'The matrix, possibly modified by rounding error, is singular or nearly singular. Any solution of a system which might be computed using the triangular factors would be meaningless'.

The *determinant* of $\mathbf{A}$ can be found as a by-product of lufac. The determinant is equal to the product of the diagonal elements of $\mathbf{U}$ multiplied by $(-1)^s$, where s is the number of row interchanges. Precautions against overflow and underflow are necessary in calculating the product; this is taken up in the exercises.

Direct Factorisation Methods

Gaussian elimination is one way to obtain triangular factors of a matrix $\mathbf{A}$. Another way is to determine the elements of the matrices $\mathbf{L} = (l_{ij})$ and $\mathbf{U} = (u_{ij})$ directly so that $\mathbf{LU} = \mathbf{A}$. There are two cases to consider, corresponding to elements above and below the diagonal.

$$i \leqslant j: \; \sum_{k=1}^{i} l_{ik} \, u_{kj} = a_{ij}$$

$$i > j: \; \sum_{k=1}^{j} l_{ik} u_{kj} = a_{ij}$$

From the first equation u_{ij} can be calculated,

$$u_{ij} = a_{ij} - \sum_{k=1}^{i-1} l_{ik} u_{kj}, \;\; i \leqslant j \tag{7.9a}$$

From the second a formula for l_{ji} can be found

$$l_{ji} = (a_{ji} - \sum_{k=1}^{i-1} l_{jk} u_{ki})/u_{ii}, \;\; j > i \tag{7.9b}$$

These equations can be used to build up the unit lower triangular matrix $\mathbf{L}$ and the upper triangular matrix $\mathbf{U}$. The elements must be calculated in a carefully defined order so that the values required on the right hand side of equations 7.9a and 7.9b are available by the time they are needed.
One possible order is

```
FOR i: = 1 TO n DO
     BEGIN
     FOR j: = i TO n DO
```

Calculate u_{ij} from equation 7.9a

FOR $j: = i + 1$ TO n DO

Calculate l_{ji} from equation 7.9b

END

The direct calculation of **L** and **U**, followed by substitution in **Ly** = **b** and **Ux** = **y**, is known as *Doolittle's method* for solving linear systems (*Crout's method* is a variant in which **U** instead of **L** is chosen to have unit diagonal elements). Doolittle's method is equivalent to Gaussian elimination; the same factors are obtained, the same amount of arithmetic is performed, and the error behaviour is the same (partial pivoting is essential for the same reason; it corresponds exactly to partial pivoting in Gaussian elimination). There are, however, two possible advantages.

(i)   On a hand calculator the inner products in equations 7.9a and 7.9b may be accumulated in a register without the need to record intermediate results (in Gaussian elimination the transcribing of intermediate matrix elements is unavoidable).

(ii)   If the inner products can be accumulated to a higher precision than the rest of the calculation without intermediate rounding, then the error of the solution can be reduced substantially (this is not possible in Pascal *per se* since only a single precision is available for real variables).

A symmetric matrix **A** is said to be *positive definite* if the scalar product $\mathbf{x}^T\mathbf{Ax}$ is positive for all non-zero vectors **x**. Positive definite matrices are non-singular and arise in many applications. It can be proved that such matrices have a factorisation of the form $\mathbf{A} = \mathbf{LL}^T$, where **L** is a lower triangular matrix with positive (but not in general unit) diagonal elements. The elements of **L** can be determined from equations similar to equations 7.9a and 7.9b; because of symmetry the amount of arithmetic is approximately halved. It can be shown that pivoting is not required for positive definite matrices. This method based on $\mathbf{LL}^T$ factorisation for positive definite systems is known as *Cholesky decomposition* and is widely used; it is efficient, stable and easy to program.

ALGOL procedures for Crout's method, Cholesky decomposition and many other computational methods of linear algebra can be found in the handbook by Wilkinson and Reinsch (1971). Comprehensive software for linear algebra problems is available in two well-known libraries, NAG and LINPACK. The latter is recognised as the leading collection of FORTRAN subroutines for the solution of non-sparse linear algebraic equations. Further information can be found in the LINPACK User's Guide by Dongarra *et al.* (1979).

7.3   SOLVING LINEAR SYSTEMS

To solve an $n \times n$ system **Ax** = **b**, we first call lufac (n, A, LU, row, eflag). If eflag is returned as 0 the matrix is non-singular; procedure lufac will have prepared the triangular factors of **A** (with row permutation) in the array LU and the permuted row indices in row. It is then only necessary to supply a right hand side **b** and compute the corresponding solution **x**, which by theorem 7.1 is unique. This may be done any number of times for different right hand sides. Since the rows of **A** are permuted in the elimination process, the components of **b** must be permuted similarly; we denote the permuted right hand side by $\mathbf{b}'$. The solution **x** is obtained from the factorised form $\mathbf{LUx} = \mathbf{b}'$ in two stages:

$$\mathbf{Ly} = \mathbf{b}', \quad \mathbf{Ux} = \mathbf{y}$$

Programming the substitution processes is easy, but first we need two new library utility routines. The first is a function to compute products without risk of underflow.

```
FUNCTION mult ( x, y : real ) :  real ;
(* Computes x*y, avoiding underflow failure. If abs(x*y)
   would be less than smallreal with a possibility of
   underflow, then mult returns 0.0 *)

   BEGIN
   IF (x = 0.0) OR (y = 0.0) THEN
      mult := 0.0
   ELSE
      IF abs(x) > 1.0 THEN
         mult := x*y
      ELSE
         IF abs(y) > smallreal/abs(x) THEN
            mult := x*y
         ELSE
            mult := 0.0   (* x*y may underflow *)
   END (* mult *)
```

Whenever there is a possibility of underflow in a product, we use the function mult instead of the operator *. This is necessary in many implementations of Pascal because underflow, even in a single term of an inner product, will cause runtime failure. Possible underflow must be detected before it can occur; a careful reading shows that mult does this.

The following function may then be used with safety to compute inner products such as those of the form

$$\sum_{j=i+1}^{n} u_{ij} x_j$$

required in the back substitution process of equations 7.4.

```
FUNCTION inprod ( VAR A : matrix ;
                  VAR x : rvector ;
                  i : range ;
                  lower, upper : range ) : real ;
(* Computes the inner product A[i,j]*x[j], j = lower,...,upper.
   Avoids underflow failure in individual products by using
   the library function mult *)

   VAR
   j   : range ;
   sum : real ;
```

```
BEGIN
sum := 0.0 ;
FOR j := lower TO upper DO
   sum := sum + mult(A[i,j], x[j]) ;
inprod := sum
END (* inprod *)
```

We now give our first library routine for solving linear systems.

```
PROCEDURE lusolv ( n   : range ;    (* order of the system    *)
               VAR LU  : matrix ;   (* triangular factors     *)
               VAR b   : rvector ;  (* right hand side vector *)
               VAR row : ivector ;  (* permuted row indices   *)
               VAR x   : rvector    (* solution vector        *)
                     ) ;

(* Computes an approximate solution of the nxn system of linear
   equations Ax = b, where b is a given vector. The routine must
   be preceded by lufac, which prepares the triangular factors of
   A in the array LU and records the permuted row indices in row.
   The system L(Ux) = b' is solved in two stages by forward and
   backward substitution, and the solution is returned in x.
   There are no error numbers in lusolv.
   Routines called from the library:  inprod (mult) *)

VAR
   i : range ;

BEGIN
(* Solve Ly = b' by forward substitution, storing the solution
   in x *)
x[1] := b[row[1]] ;
FOR i := 2 TO n DO
   x[i] := b[row[i]] - inprod (LU, x, row[i], 1, i-1) ;

(* Solve Ux = y by back substitution *)
x[n] := x[n] / LU[row[n],n] ;
FOR i := n-1 DOWNTO 1 DO
   x[i] := (x[i] - inprod (LU, x, row[i], i+1, n))
                                         / LU[row[i],i]
END (* lusolv *)
```

Notes

(i)   inprod (mult) in the list of routines called from the library indicates that inprod is called directly and this in turn calls mult (nesting to any depth is possible).

(ii)   The components of b are not permuted explicitly; they are addressed through the vector row produced by lufac.

The solution of $Ax = b$ computed by lusolv is usually not correct to full machine precision because of rounding errors in the calculation, particularly in lufac. Indeed, it is possible for the computed solution to contain a substantial error. We shall see in section 7.4 that this depends on a property of the coefficient matrix called the *condition number* which in general is not easy to determine. At this stage there is no way to assess the accuracy of the solution. This may be tolerable in many applications; there are often data errors due to linearising assumptions in the original problem, and the user generally has no reason to be concerned if additional errors are introduced in the computation. However, we shall next see how to improve the solution from lusolv virtually to full machine accuracy at little extra cost, and at the same time estimate the sensitivity of the solution to data errors.

Iterative Refinement

Suppose we have computed an approximate solution $x^{(1)}$ of $Ax = b$. Some or all of the components of $x^{(1)}$ contain errors, usually in the last few significant digits. Since $x^{(1)}$ is not the exact solution of $Ax = b$, the *residual vector* defined by

$$r^{(1)} = b - Ax^{(1)}$$

will be non-zero. Let us set up and solve the system

$$Ad^{(1)} = r^{(1)}$$

where $d^{(1)}$ is a new vector of unknowns. If we could compute $d^{(1)}$ exactly we could obtain the exact solution of $Ax = b$; the solution would be given by $x = x^{(1)} + d^{(1)}$, since

$$Ad^{(1)} = r^{(1)} = b - Ax^{(1)}$$

and so

$$A(x^{(1)} + d^{(1)}) = b$$

We cannot compute $d^{(1)}$ exactly for the same reason that we cannot compute x exactly in the first place: there are unavoidable rounding errors in lufac and lusolv. Under suitable conditions, however, the computed $d^{(1)}$ may be taken as a correction to at least some of the incorrect digits in the components of $x^{(1)}$, and we obtain an improved approximation

$$x^{(2)} = x^{(1)} + d^{(1)}$$

The residual vector $r^{(2)} = b - Ax^{(2)}$ can now be calculated, and the whole cycle may be repeated. We have an iterative scheme

$$\left.\begin{array}{l} r^{(k)} = b - Ax^{(k)} \\ d^{(k)} = A^{-1}r^{(k)} \\ x^{(k+1)} = x^{(k)} + d^{(k)} \\ k = 1, 2, \ldots \end{array}\right\} \tag{7.10}$$

This is known as *iterative refinement*.

There are three points which are fundamental to understanding and implementing 7.10.

(i)   Systems of the form $Ad = r$ are to be solved for a succession of different right hand sides. Since the coefficient matrix is the same in all cases, the factorisation of **A** obtained in lufac can be used repeatedly. The only additional work is in lusolv, and this is much less than the initial work of factorisation.

(ii)   The components of the residual vector $\mathbf{r}^{(k)}$ are given by

$$r_i^{(k)} = b_i - \sum_{j=1}^{n} a_{ij} x_j^{(k)}, \quad i = 1, \ldots, n$$

These values will be small, of the order of the least significant digits of b_i. Since the residuals are formed as differences of much larger terms, it is essential to accumulate the inner product and perform the final subtraction using *higher precision*, that is, more digits than normal. If ordinary precision is used the values obtained for the $r_i^{(k)}$ will be largely meaningless.

(iii)   The iteration is terminated when the correction $\mathbf{d}^{(k)}$ is sufficiently small that the largest component of $\mathbf{x}^{(k)}$ is effectively unchanged in ordinary precision (we shall use tolmin to test for this). The largest component of $\mathbf{x}^{(k)}$, at least, is then correct to full machine accuracy.

An analysis of convergence for iterative refinement, with simplifying assumptions, is given by Forsythe and Moler (1967). There are matrices which are so ill-conditioned that iterative refinement fails to converge, but these are very rare. In practice the refinement process nearly always terminates in less than five steps.

In order to write an accurate linear equation solver using iterative refinement, we need a routine to compute accurate residuals as pointed out in (ii) above. This presents a problem, since in Pascal only one precision of real variables is available. Most FORTRAN implementations, however, provide DOUBLE PRECISION variables with about twice the normal number of significant digits. For this single purpose we shall use a routine called RESIDU written in FORTRAN.

A FORTRAN routine can be called from a Pascal program provided the appropriate declaration is made; residu is declared in the mathlib library as

```
PROCEDURE residu (      n    : range ;
                   VAR A     : matrix ;
                   VAR x, b  : rvector ;
                   VAR r     : rvector ) ;
(* An external FORTRAN subroutine to compute accurately the
   residual vector r = b - Ax, given an nxn matrix A and vectors
   x, b of order n. The dimensions of the arrays must be equal to
   upbnd (currently 100). The calculation of inner products and
   final subtraction should be performed in double precision *)

   FORTRAN ;
```

The statement FORTRAN in place of the procedure body signifies that procedure residu corresponds to a FORTRAN subroutine in an external file. The file containing the compiled code of RESIDU must be loaded with mathlib when the library is to be used; the arrangements for this are, of course, system-dependent. This special requirement is emphasised in a note at the head of the library (see appendix B).

RESIDU could be written in standard FORTRAN as follows.

```
      SUBROUTINE RESIDU (N, A, X, B, R)
      DIMENSION  A(100,100), X(100), B(100), R(100)
C        FORTRAN subroutine to compute the vector
C        R = B - AX of order N <= 100 using double
C        precision accumulation of inner products
C
      DOUBLE PRECISION  SUM, AIJ, XJ
      DO 2  I = 1, N
        SUM = B(I)
        DO 1  J = 1, N
          AIJ = A(J,I)
          XJ  = X(J)
          SUM = SUM - AIJ*XJ
    1   CONTINUE
        R(I) = SUM
    2 CONTINUE
      RETURN
      END
```

Notes

(i)   Care should be taken in interfacing Pascal and FORTRAN that the precision used for real variables is the same in both language implementations. The availability of double precision, and the appropriate syntax, should be checked in the system documentation.

(ii)   Different mapping conventions are used for arrays in Pascal and FORTRAN. The array element A(J, I) in RESIDU corresponds to A[i, j] in Pascal.

Using lusolv and residu we can now write our library routine for the accurate solution of linear systems.

```
PROCEDURE linsolv ( n    : range ;    (* order of the system        *)
                    VAR A    : matrix ;   (* nxn coefficient matrix  *)
                    VAR LU   : matrix ;   (* triangular factors of A *)
                    VAR b    : rvector ;  (* right hand side vector  *)
                    VAR row : ivector ;   (* permuted row indices    *)
                    VAR x    : rvector ;  (* solution vector         *)
                    VAR eflag : integer ) ;

(* Computes an accurate solution of the nxn system of linear
   equations Ax = b, where b is a given vector. The routine must
   be preceded by lufac, which prepares the triangular factors of
   A in the array LU and records the permuted row indices in row.
   An approximate solution is computed by substitution; this is
   then improved by iterative refinement, and the solution
   correct to machine accuracy is returned in x.
   The error number for eflag is:
     =1  Iterative refinement fails to converge - matrix is too
         ill-conditioned for the system to be solved  (fatal).
   Routines called from the library:
   lusolv (inprod (mult)), tolmin, errormessage.
   **NOTE: an external FORTRAN subroutine RESIDU is called to
   compute accurate residuals - see comment at head of library *)

CONST
    maxiter = 20 ;    (* maximum number of iterations allowed *)

VAR
    xmax, dmax : real ;
    count      : integer ;
    i          : range ;
    r, d       : rvector ;
```

```
BEGIN
lusolv (n, LU, b, row, x) ;   (* compute approximate
                                   solution x *)
count := 0 ;
REPEAT   (* iterative refinement *)
    residu (n, A, x, b, r) ;       (* compute r = b - Ax *)
    lusolv (n, LU, r, row, d) ;   (* solve LU.d = r *)
    (* Add d to x as correction and compute maximum components
       of x and d *)
    xmax := 0.0 ;   dmax := 0.0 ;
    FOR i := 1 TO n DO
        BEGIN
        x[i] := x[i] + d[i] ;
        IF xmax < abs(x[i]) THEN
            xmax := abs(x[i]) ;
        IF dmax < abs(d[i]) THEN
            dmax := abs(d[i])
        END ;
    count := count + 1
UNTIL (dmax < tolmin(xmax))   (* convergence achieved *)
    OR (count = maxiter) ;      (* maximum iterations used *)

IF dmax < tolmin(xmax) THEN
    eflag := 0   (* solution is correct to machine accuracy *)
ELSE
    errormessage ('linsolv ', 1,
        'refinement fails to converge  ', fatal, eflag)

END (* linsolv *)
```

An Example Program

We illustrate the solution of a system of linear equations using lufac, lusolv and
linsolv by means of the following program. This reads the order n of a system,
the n x n matrix of coefficients, and the n-component vector of right hand sides.
It outputs the approximate solution computed by lusolv and the accurate
solution computed by linsolv.

```
PROGRAM testlineq ( input, output ) ;
(* This is an example program to illustrate the solution of an
   nxn system of linear equations Ax = b, where b is a given
   vector: n, A and b are read in as data, lufac is called to
   obtain triangular factors L and U of A, then lusolv is used to
   compute an approximate solution of LU.x = b by substitution.
   Finally linsolv is used to compute an accurate solution by
   iterative refinement *)

TYPE
    range   = 1..100 ;
    matrix  = ARRAY [range, range] OF real ;
    rvector = ARRAY [range] OF real ;
    ivector = ARRAY [range] OF integer ;

VAR
    i, j, n : range ;
    eflag   : integer ;
    A, LU   : matrix ;
    b, x    : rvector ;
    row     : ivector ;


PROCEDURE lufac ( n : range ;  VAR A, LU : matrix ;
                  VAR row : ivector ;  VAR eflag : integer ) ;
    EXTERN ;
```

```
PROCEDURE lusolv ( n : range ;   VAR LU : matrix ;
                   VAR b : rvector ;   VAR row : ivector ;
                   VAR x : rvector ) ;
   EXTERN ;


PROCEDURE linsolv ( n : range ;   VAR A, LU : matrix ;
                    VAR b : rvector ;   VAR row : ivector ;
                    VAR x : rvector ;   VAR eflag : integer ) ;
   EXTERN ;


BEGIN
read (n) ;
FOR i := 1 TO n DO
   FOR j := 1 TO n DO
      read (A[i,j]) ;
FOR i := 1 TO n DO
   read (b[i]) ;

eflag := 0 ;
lufac (n, A, LU, row, eflag) ;
IF eflag = 0 THEN
   BEGIN
   lusolv (n, LU, b, row, x) ;
   writeln ;
   writeln ('   approximate solution') ;   writeln ;
   FOR i := 1 TO n DO
      writeln (x[i]:20:14) ;
   linsolv (n, A, LU, b, row, x, eflag) ;
   IF eflag = 0 THEN
      BEGIN
      writeln ;   writeln ;
      writeln ('   accurate solution') ;   writeln ;
      FOR i := 1 TO n DO
         writeln (x[i]:20:14)
      END
   END
END .
```

Example 7.5

Solve the linear system with the following 7 x 7 coefficient matrix and right hand
side.

360360.0	180180.0	120120.0	90090.0	72072.0	60060.0	51480.0
180180.0	120120.0	90090.0	72072.0	60060.0	51480.0	45045.0
120120.0	90090.0	72072.0	60060.0	51480.0	45045.0	40040.0
90090.0	72072.0	60060.0	51480.0	45045.0	40040.0	36036.0
72072.0	60060.0	51480.0	45045.0	40040.0	36036.0	32760.0
60060.0	51480.0	45045.0	40040.0	36036.0	32760.0	30030.0
51480.0	45045.0	40040.0	36036.0	32760.0	30030.0	27720.0
273702.0	131703.0	88517.0	67639.0	55211.0	46891.0	40889.0

The matrix is highly ill-conditioned, a property which will be discussed in
section 7.4. One effect is that rounding errors in the calculation are greatly
amplified in the solution†. The right hand side is chosen so that the exact solu-

†The matrix elements are arranged to be whole numbers lying within the range
of machine integers so that there is no machine rounding in representing the data.
If we overlook this we introduce data errors, and the true solution is then substantially different from the one expected (it is in fact machine-dependent).

tion is known to be $x = (1, -1, 1, -1, 1, -1, 1)^T$. The output from program testlineq is

```
approximate solution

  1.00000000000081
 -1.00000000002983
  1.00000000027231
 -1.00000000101978
  1.00000000181889
 -1.0000000153903
  1.00000000049702


accurate solution

  1.00000000000000
 -1.00000000000000
  1.00000000000000
 -1.00000000000000
  1.00000000000000
 -1.00000000000000
  1.00000000000000
```

Iterative refinement converges to the exact solution in two steps. The approximate solution from lusolv computed on the Amdahl V/7, although inaccurate in all components, is quite good; on a computer with lower precision the errors would be larger. We shall see in the next section that the difference between the approximate and accurate solutions can be used to estimate the condition number of the matrix; this will finally tell us how much trust to put in the computed solution. □

Some mathematical libraries have routines to compute the inverse of a non-singular matrix. This problem can be solved using the ideas and procedures already developed in this chapter; as the explicit matrix inverse is not often required, the details are left to the exercises.

We conclude this section with a table showing the performance of our routines on linear systems of orders 20, 40, . . ., 100 with randomly generated matrix elements. The statistic quoted is the central processor time in seconds (to the nearest 1/10 second) on the Amdahl V/7.

Order of system	lufac	lusolv	linsolv
20	0.5	0.1	0.1
40	1.2	0.1	0.3
60	2.9	0.2	0.5
80	5.6	0.2	0.9
100	9.7	0.3	1.2

The table reflects our theoretical results on n^3 time dependence for elimination and n^2 time dependence for substitution. The highly practical effect is that most of the computation time is spent in lufac.

7.4   CONDITION AND NORMS

The problem of solving a system of linear equations, like any numerical problem, is said to be *ill-conditioned* if a small perturbation of the data (in this case the matrix elements and right hand side) can produce a large change in the solution. Some idea of the condition is necessary if we are to assess the worth of a computed solution. We shall see that the condition of a linear system is determined by the coefficient matrix.

Example 7.6

Consider the system

$$4.0x + 5.8y = 9.8$$
$$2.0x + 3.0y = 5.0$$

We note first that an increase of about three per cent in the coefficient 5.8, changing it to 6.0, will produce the matrix of example 7.2. This matrix is singular, and the equations would then be inconsistent with no solution. Equivalently, we can say that the rows (and columns) of the coefficient matrix are almost linearly dependent, that is, one row (or column) is almost a multiple of the other.

Gaussian elimination would lead to

$$4.0x + 5.8y = 9.8$$
$$0.1y = 0.1$$

with a diagonal element 0.1 which, significantly, is much smaller than the pivot 4.0. The exact solution is $x = y = 1.0$. We can interpret this graphically as follows.

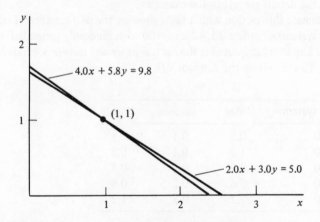

Figure 7.2   Solution of a 2×2 ill-conditioned system

The straight lines are almost parallel because the matrix is almost singular (if the matrix were exactly singular the lines would be parallel). It is clear from the graph that a small change in the coefficients, corresponding to a small change in the gradients of the lines, will produce a large shift in the point of intersection. For example, changing the coefficient 3.0 to 3.1 we find

$$4.0x + 5.8y = 9.8$$
$$0.2y = 0.1$$

and the exact solution becomes $x = 1.725, y = 0.5$. Conversely, if we retain the original system and substitute the grossly inaccurate 'solution' $(1.725, 0.5)^T$, we find that the *residuals* are

$$9.8 - 4.0 \times 1.725 - 5.8 \times 0.5 = 0.0$$
$$5.0 - 2.0 \times 1.725 - 3.0 \times 0.5 = 0.05$$

The solution is in error by 50–75 per cent, but the residuals are at most one per cent of the right hand side. The system appears to be almost exactly satisfied; this is typical of ill-conditioned problems. □

Rounding errors in a calculation effectively perturb the data (see, for example, Forsythe and Moler (1967), section 20). If the problem is ill-conditioned, this means that the computed solution is likely to be a poor approximation to the true solution. Iterative refinement with higher-precision calculation of residuals can virtually eliminate this computational error. The results, however, may be misleading: there are usually inherent errors in the data (often errors of experimental measurement), and there is little point in obtaining a solution which is computationally almost perfect if the accuracy of the data does not warrant it. We need a quantitative measure of the condition of a linear system and an analysis of how data errors can affect the solution.

Vector and Matrix Norms

Norms are non-negative real numbers which provide a measure of the size of (among other mathematical objects) vectors and matrices. A norm of a vector $\mathbf{x}$ is denoted by $\|\mathbf{x}\|$, often with a subscript to indicate a particular type of norm. For example, the ordinary geometric length of a vector is a norm: if $\mathbf{x} = (1, 2, -2)^T$ then $\|\mathbf{x}\|_2 = (1^2 + 2^2 + (-2)^2)^{1/2} = 3$.

In general, a *vector norm* is an assignment of a real number to each n-component vector, with the following properties

$$\left. \begin{array}{l} \|\mathbf{x}\| > 0 \text{ if } \mathbf{x} \neq \mathbf{0}, \quad \|\mathbf{0}\| = 0 \\ \|\lambda \mathbf{x}\| = |\lambda| \cdot \|\mathbf{x}\| \text{ for any scalar } \lambda \\ \|\mathbf{x} + \mathbf{y}\| \leqslant \|\mathbf{x}\| + \|\mathbf{y}\| \text{ (triangle inequality)} \end{array} \right\} \tag{7.11}$$

These requirements are satisfied (indeed suggested) by the *euclidean norm*

$$\|\mathbf{x}\|_2 = \left(\sum_{i=1}^{n} |x_i|^2 \right)^{1/2} = (\mathbf{x}^T \mathbf{x})^{1/2}$$

This corresponds to the ordinary length of a vector as in the example above. A different type of norm which is useful in applications is the *maximum norm*

$$\|\mathbf{x}\|_\infty = \max_{1 \leqslant i \leqslant n} |x_i|$$

If $\mathbf{x} = (1, 2, -2)^\mathrm{T}$ then $\|\mathbf{x}\|_\infty = 2$. It is easy to verify that this norm also satisfies the requirements 7.11.

We now define a norm of an $n \times n$ matrix $\mathbf{A}$ by

$$\|\mathbf{A}\| = \max_{\mathbf{x} \neq \mathbf{0}} \frac{\|\mathbf{A}\mathbf{x}\|}{\|\mathbf{x}\|} \tag{7.12}$$

where the maximum is taken over all n-component vectors $\mathbf{x}$ excluding the zero vector†. Since $\mathbf{A}\mathbf{x}$ and $\mathbf{x}$ are vectors and $\mathbf{x} \neq \mathbf{0}$, we can form the ratio of vector norms (euclidean, maximum or other) required on the right hand side. The matrix norm so defined is said to be *subordinate* to the corresponding vector norm and is denoted by the same subscript. It can be shown from the definition 7.12 that any subordinate matrix norm has properties analogous to 7.11, and also

$$\|\mathbf{A}\mathbf{B}\| \leqslant \|\mathbf{A}\| \cdot \|\mathbf{B}\|$$

for all $n \times n$ matrices. If $\mathbf{x}^*$ is a vector for which the maximum in equation 7.12 is attained, then for any vector $\mathbf{x} \neq \mathbf{0}$

$$\frac{\|\mathbf{A}\mathbf{x}\|}{\|\mathbf{x}\|} \leqslant \frac{\|\mathbf{A}\mathbf{x}^*\|}{\|\mathbf{x}^*\|} = \|\mathbf{A}\|$$

and we have the important result

$$\|\mathbf{A}\mathbf{x}\| \leqslant \|\mathbf{A}\| \cdot \|\mathbf{x}\| \tag{7.13}$$

The particular advantage of the maximum vector norm lies in the fact that the subordinate matrix norm can be expressed by an easily computed formula. It can be shown from equation 7.12 that

$$\|\mathbf{A}\|_\infty = \max_{1 \leqslant i \leqslant n} \sum_{j=1}^{n} |a_{ij}|$$

This is the maximum row sum of absolute values of elements in the matrix $\mathbf{A}$. For example, if

$$\mathbf{A} = \begin{bmatrix} 1 & -2 & 0 \\ 2 & -1 & 1 \\ 1 & 0 & 1 \end{bmatrix}$$

then

$$\|\mathbf{A}\|_\infty = \max (3, 4, 2) = 4$$

†It can be shown that the maximum exists and is attained for some n-component vector $\mathbf{x}^*$.

We shall use these results, in particular the triangle inequality and inequality 7.13, to quantify the discussion of the condition of a linear system.

Condition of a Linear System

Suppose that the coefficients and right hand side in $Ax = b$ are perturbed by amounts δA and δb respectively (these symbols denote a matrix and a vector). How is the solution x affected? We have for the perturbed system (assuming that $A + \delta A$ is non-singular),

$$(A + \delta A)(x + \delta x) = b + \delta b$$

hence to first order in small quantities†

$$A \cdot \delta x + \delta A \cdot x \approx \delta b$$

If A is non-singular

$$\delta x \approx A^{-1}\,\delta b - A^{-1} \cdot \delta A \cdot x$$

Taking norms and using the triangle inequality and 7.13 we find

$$\|\delta x\| \lesssim \|A^{-1}\| \, (\|\delta b\| + \|\delta A\| \cdot \|x\|)$$

If $b \neq 0$ (so $x \neq 0$) we can write this as

$$\frac{\|\delta x\|}{\|x\|} \lesssim \|A^{-1}\| \left(\frac{\|\delta b\|}{\|x\|} + \|\delta A\| \right)$$

Now taking norms in $Ax = b$ we have $\|b\| \leqslant \|A\| \cdot \|x\|$, hence

$$\frac{1}{\|x\|} \leqslant \frac{\|A\|}{\|b\|}$$

Using this in the right hand side above we find

$$\frac{\|\delta x\|}{\|x\|} \lesssim \|A\| \cdot \|A^{-1}\| \left(\frac{\|\delta b\|}{\|b\|} + \frac{\|\delta A\|}{\|A\|} \right) \tag{7.14}$$

This is an approximate bound for the relative perturbation of the solution x of a system $Ax = b$ induced by perturbations in the data A and b. The factor on the right of inequality 7.14

$$\text{cond}(A) = \|A\| \cdot \|A^{-1}\|$$

is called the *condition number* of A. The smallest possible value of the condition number in a subordinate norm is 1 (see exercises). If cond (A) is not too large, data errors and rounding errors in the calculation (using a stable algorithm such

---

†It is assumed in this approximate treatment that $\|A^{-1}\| \times \|\delta A\| \ll 1$. An exact treatment without this assumption can be given, leading to a more complicated formula (see, for example, Phillips and Taylor (1973), chapter 9).

as Gaussian elimination with pivoting) will not affect the solution unduly. If, on the other hand, cond($\mathbf{A}$) is large, data errors may be magnified and produce a large error in the solution irrespective of the computational method employed.

It can be shown in a similar way that if $\mathbf{x}$ is the exact solution of $\mathbf{A}\mathbf{x} = \mathbf{b}$ and $\overline{\mathbf{x}}$ is an incorrect solution with residual vector $\mathbf{r}$, then

$$\frac{\|\overline{\mathbf{x}} - \mathbf{x}\|}{\|\mathbf{x}\|} \leqslant \text{cond}(\mathbf{A}) \cdot \frac{\|\mathbf{r}\|}{\|\mathbf{b}\|}$$

In practice, having computed a solution $\overline{\mathbf{x}}$, we can find the residual vector $\mathbf{r}$. However, contrary to intuition, small relative residuals do *not* necessarily indicate an accurate solution if the problem is ill-conditioned; this is reflected in the factor cond($\mathbf{A}$) above. In example 7.6 the residual vector was found to be $(0.0, 0.05)^{\mathrm{T}}$ for $\overline{\mathbf{x}} = (1.725, 0.5)^{\mathrm{T}}$ and $\mathbf{b} = (9.8, 5.0)^{\mathrm{T}}$. The condition number of the matrix is about 200 in the maximum norm. It follows that the possible relative error of $\overline{\mathbf{x}}$ is $200 \times 0.05/9.8 \approx 1.0$. The error could be as much as 100 per cent; in fact it is over 70 per cent.

We now return to the approximate solution $\overline{\mathbf{x}}$ and the accurate solution $\mathbf{x}$ of $\mathbf{A}\mathbf{x} = \mathbf{b}$ obtained by the routines lusolv and linsolv in example 7.5. The reason for the inaccuracy of the approximate solution is rounding error in Gaussian elimination magnified by the ill-conditioning of the matrix†. The bound on the error $\|\overline{\mathbf{x}} - \mathbf{x}\|$ depends directly on the condition number and also on the elementary rounding error of the computer (the relative precision u). It is shown by Forsythe and Moler (1967) that the relation between these quantities is (roughly)

$$\text{cond}(\mathbf{A}) \sim \frac{1}{nu} \cdot \frac{\|\overline{\mathbf{x}} - \mathbf{x}\|}{\|\mathbf{x}\|}$$

This can be used to obtain a rough estimate of the condition number.

In the computer output of example 7.5 we have

$$\|\overline{\mathbf{x}} - \mathbf{x}\|_\infty = 0.18 \times 10^{-8}, \quad \|\mathbf{x}\|_\infty = 1$$

hence

$$\text{cond}(\mathbf{A}) \sim \frac{1}{7 \times 0.2 \times 10^{-15}} \times \frac{0.18 \times 10^{-8}}{1} \approx 0.13 \times 10^7$$

This shows that the matrix is very ill-conditioned, although the true condition number may differ from the estimate by a factor of 10 or more. Now if the data on the right hand side in example 7.5 were not exact but only rounded to the nearest whole number, there would be a possible error of ± 0.5 in the components of $\mathbf{b}$, hence

† In *inverse error analysis* the computed solution $\overline{\mathbf{x}}$ is regarded as the exact solution of a perturbed system $(\mathbf{A} + \delta\mathbf{A})\,\overline{\mathbf{x}} = \mathbf{b}$. A bound or estimate is obtained for $\|\delta\mathbf{A}\|/\|\mathbf{A}\|$, and this can be used in inequality 7.14 if an estimate is known for cond($\mathbf{A}$). This has been developed in depth by Wilkinson (1963).

$$\frac{\|\delta b\|_\infty}{\|b\|_\infty} \leqslant \frac{0.5}{274000} \approx 0.18 \times 10^{-5}$$

Then by inequality 7.14, using the estimate of cond(A) above, we find

$$\frac{\|\delta x\|_\infty}{\|x\|_\infty} \lesssim 0.13 \times 10^7 \times 0.18 \times 10^{-5} = 2.3$$

This would mean that (under our assumption regarding the data error) no figures even in the 'accurate' solution could be claimed correct! This startling result is due to the extreme ill-conditioning of the coefficient matrix.

Least-squares Approximation

To see how an ill-conditioned system of linear equations can arise, we consider the problem of fitting a polynomial $a_0 + a_1 x + \ldots + a_{n-1} x^{n-1}$ to a set of data points (x_i, f_i), $i = 1, \ldots, m$, representing the measured values of some quantity $f(x)$. If the polynomial is to pass through all the data points, there are m conditions to be satisfied

$$a_0 + a_1 x_1 + \ldots + a_{n-1} x_1^{n-1} = f_1$$
$$\cdot \quad \cdot \quad \cdot \quad \cdot \quad \cdot \quad \cdot \quad \cdot \quad \cdot \quad \cdot \quad \cdot \quad \cdot$$
$$\cdot \quad \cdot \quad \cdot \quad \cdot \quad \cdot \quad \cdot \quad \cdot \quad \cdot \quad \cdot \quad \cdot$$
$$a_0 + a_1 x_m + \ldots + a_{n-1} x_m^{n-1} = f_m$$

This is an $m \times n$ system of linear equations in the unknowns $a_0, a_1, \ldots, a_{n-1}$ (all the other quantities are given).

If $m = n$ the system can be solved in the usual way, provided that the x_i are all distinct; the result is the interpolating polynomial for the data points (see section 3.5). Usually, however, to reduce the effects of experimental error, measurements of f are made at a large number of different points, so that $m \gg n$. The system is then said to be *overdetermined*. Any subset of n of the equations can be solved, but the resulting solution in general does not satisfy the remaining $m - n$ equations. In this case we seek a solution which best satisfies the system as a whole: no single equation is expected to be satisfied exactly, but all m equations are to be satisfied approximately in some 'best' sense.

A commonly used criterion for best approximation is the so-called *principle of least squares*. For an $m \times n$ system $Ca = d$ with $m > n$, a *least-squares solution* is an n-component vector a^* which minimises the sum of squares of residuals

$$\sum_{i=1}^{m} (d_i - (Ca)_i)^2$$

Thus the euclidean norm $\|r^*\|_2 = \|d - Ca^*\|_2$ is as small as possible, though usually greater than zero.

Let **a** be any n-component vector, and consider the residual vector

$$\mathbf{r} = \mathbf{d} - \mathbf{Ca} = \mathbf{d} - \mathbf{Ca^*} + \mathbf{Ca^*} - \mathbf{Ca}$$
$$= \mathbf{r^*} + \mathbf{C(a^* - a)}$$

Premultiplying by the transpose and using $\mathbf{r}^T\mathbf{r} = \|\mathbf{r}\|_2{}^2$, we find

$$\|\mathbf{r}\|_2{}^2 = \|\mathbf{r^*}\|_2{}^2 + \mathbf{r^{*T}C(a^* - a)}$$
$$+ (\mathbf{r^{*T}C(a^* - a)})^T + \|\mathbf{C(a^* - a)}\|_2{}^2$$

It follows that $\|\mathbf{r}\|_2 \geqslant \|\mathbf{r^*}\|_2$, so that $\mathbf{a^*}$ is a least-squares solution, provided that the inner product $(\mathbf{C}^T\mathbf{r^*})^T(\mathbf{a^* - a}) = 0$ for all **a**. This condition is satisfied if $\mathbf{C}^T\mathbf{r^*}$ is the zero vector, that is $\mathbf{C}^T\mathbf{d} - \mathbf{C}^T\mathbf{Ca^*} = \mathbf{0}$.

The condition $\mathbf{C}^T\mathbf{Ca} = \mathbf{C}^T\mathbf{d}$ represents an $n \times n$ system of linear equations known as the *normal equations* of the least-squares problem. It can be shown that if the columns of **C** are linearly independent†, then the $n \times n$ symmetric matrix $\mathbf{C}^T\mathbf{C}$ is non-singular, indeed positive definite. The normal equations then have a unique solution $\mathbf{a^*}$ which can be determined by standard methods. The simplest case is to fit a straight line $y = a_0 + a_1 x$ to a set of data points $(x_1, f_1), \ldots, (x_m, f_m)$. In this case

$$C = \begin{bmatrix} 1 & x_1 \\ \cdot & \cdot \\ \cdot & \cdot \\ \cdot & \cdot \\ 1 & x_m \end{bmatrix}, \mathbf{a} = \begin{bmatrix} a_0 \\ a_1 \end{bmatrix}, \mathbf{d} = \begin{bmatrix} f_1 \\ \cdot \\ \cdot \\ \cdot \\ f_m \end{bmatrix}$$

and the normal equations are

$$\begin{bmatrix} m & \Sigma x_i \\ \Sigma x_i & \Sigma x_i^2 \end{bmatrix} \begin{bmatrix} a_0 \\ a_1 \end{bmatrix} = \begin{bmatrix} \Sigma f_i \\ \Sigma x_i f_i \end{bmatrix}$$

where all summations are from 1 to m. These may be recognised as the equations for linear regression analysis in statistics. The equations can be solved by elimination to give the desired coefficients a_0 and a_1.

Although the normal equations for polynomial curve-fitting are often set up and solved in this way, it should be noted that for larger n the columns of **C** are almost linearly dependent; for $n > 3$ the normal equations become very ill-conditioned. We should either use low-degree polynomials, possibly approximating different sections of the data with different curves (*piecewise* polynomial approximation), or else use methods based on an alternative approach (see, for example, Phillips and Taylor (1973), chapter 5).

The least-squares criterion can also be used to approximate a function $f(x)$ on a closed interval rather than on a discrete data set. For the interval $[0, 1]$ and an approximating polynomial

† that is, no column can be expressed as a linear combination of the other columns.

$$\sum_{i=0}^{n-1} a_i x^i$$

it is straightforward to derive the normal equations

$$\sum_{j=0}^{n-1} h_{ij}a_j = g_i, \quad i = 0, 1, \ldots, n-1$$

where

$$h_{ij} = \int_0^1 x^{i+j}\, dx = \frac{1}{i+j+1}$$

and

$$g_i = \int_0^1 x^i f(x)\, dx$$

(these correspond to the discrete case with summations replaced by integrals).
The $n \times n$ matrix $H_n = (h_{ij})$, which takes the form

$$H_n = \begin{bmatrix} 1 & 1/2 & 1/3 & \cdot & & \cdot \\ 1/2 & 1/3 & 1/4 & & & \cdot \\ 1/3 & 1/4 & 1/5 & & & \cdot \\ \cdot & & & \cdot & & \cdot \\ \cdot & & & & \cdot & \\ \cdot & \cdot & \cdot & \cdot & \cdot & \dfrac{1}{2n-1} \end{bmatrix}$$

is the celebrated *Hilbert matrix* of order n. For $n > 3$ the Hilbert matrices are
very ill-conditioned. The matrix used in example 7.5 is $360{,}360 \times H_7$, and this,
like H_7, has a condition number in excess of 10^8.

7.5   LARGE LINEAR SYSTEMS

We recall from the table at the end of section 7.3 that our elimination and
substitution routines can solve a dense 100×100 system in about 10 seconds on
the Amdahl V/7. From the n^3 time dependence of the elimination process, we
might expect to be able to solve a dense 1000×1000 system in about $10^3 \times 10$
seconds (say three hours of central processor time). If time were the only
consideration this would be within the bounds of possibility. There is, however,
also a space limitation. At least one $n \times n$ array is used, and for $n = 1000$ this
would require 10^6 storage locations, well beyond the main storage capacity
available on most present-day computers.

It is nevertheless possible to solve dense systems of this size by elimination.
The matrix elements are held in secondary store, for example on magnetic tape,
and only a fraction of the rows (or columns) are brought into core for processing
at any one time. It is essential to keep the number of memory transfers between

main and secondary store to a minimum. Gaussian elimination as we have
described it is not suitable for this purpose, but there are ways to organise the
elimination process so that this can be done (see Dahlquist and Björck (1974),
chapter 5, for an overview).

Large linear systems occurring in many important applications are *sparse*, that
is, most of the matrix elements are zero. Sparse systems typically arise in discretisation methods for solving boundary value problems in ordinary and partial
differential equations. A common structure is a *band matrix* in which the non-
zero elements are concentrated in a band about the main diagonal: an element
a_{ij} can be non-zero only if $-q \leqslant i - j \leqslant p$, where p and q are parameters of the
matrix. Such a matrix is said to have a *band width* of $w = p + q + 1$. For example,
the matrix

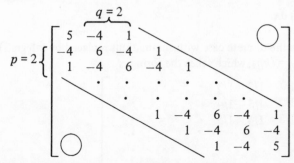

which occurs in the numerical solution of problems in elasticity theory, is a band
matrix with $p = q = 2$ and $w = 5$.

A band matrix can be stored compactly in a rectangular array of n rows and
w columns†. If $w \ll n$, it is possible to handle systems up to a high order before
the storage limit is encountered. Band systems can be solved by elimination or
factorisation methods with modifications to cater for the special data structure.
The important point is that elements lying outside the band are known to be zero
and can therefore be ignored. The effect of pivoting is to extend the band
marginally to the right. We see from the example that if rows 1 and 3 are interchanged, two extra non-zero elements appear on the right of the first row.
In general, the maximum expansion is an extra p diagonals to the right of the
band. If the matrix is positive definite, pivoting is not required and no expansion
takes place. In any case the amount of arithmetic is reduced in comparison with
the solution of a full $n \times n$ system by a factor of about $(w/n)^2$.

Band matrices of the particularly simple form with $p = q = 1$ are called *tridiagonal*. Systems with positive definite coefficient matrices of tridiagonal form
occur frequently; these can be solved by elimination and substitution in a total
of about $8n$ arithmetic operations. The reader is referred to Wilkinson and
Reinsch (1971) for procedures to solve general and positive definite band systems
and for a discussion of appropriate data structures.

†The band structure depends on the ordering of the equations and unknowns. In
some applications it is a non-trivial problem to devise an ordering which will
produce a reasonably small band width. Considerable work has been done in this
area in recent years (see George and Liu (1981)).

Application: Boundary Value Problems

To see how a sparse system of linear equations can arise, we consider the problem of determining the temperature profile across an infinite conducting bar of unit square cross-section when a specified temperature distribution is maintained around the boundary.

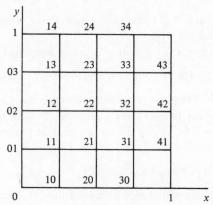

Figure 7.3 Cross-section of a square bar with superimposed grid

The steady-state temperature $u(x, y)$ is governed by *Laplace's equation* in two dimensions

$$\frac{\partial^2 u}{\partial x^2} + \frac{\partial^2 u}{\partial y^2} = 0$$

The values of $u(x, y)$ are given on the boundary of the square region; these are the *boundary values*. The problem is to determine $u(x, y)$ in the interior†.

The problem can be discretised as follows. A quantity $h = 1/m$ is defined, and a square grid of *mesh size h* is superimposed on the region (in figure 7.3 $h = 1/4$). We now have a finite set of *mesh points* $(x_i, y_j) = (ih, jh)$, $0 \leqslant i, j \leqslant m$. The values $u(x_i, y_j)$ are known on the boundary and we wish to determine the values at the interior points (in the figure there are nine of these).

The basic idea is to replace the partial derivatives in Laplace's equation by *central difference quotients*, which give an approximation of order h^2.

$$\left(\frac{\partial^2 u}{\partial x^2}\right)_{i,j} = \frac{u(x_{i+1}, y_j) - 2u(x_i, y_j) + u(x_{i-1}, y_j)}{h^2} + 0(h^2)$$

$$\left(\frac{\partial^2 u}{\partial y^2}\right)_{i,j} = \frac{u(x_i, y_{j+1}) - 2u(x_i, y_j) + u(x_i, y_{j-1})}{h^2} + 0(h^2)$$

(these equations can be verified by Taylor's series expansion about (x_i, y_j) on

†This is a model problem; more complicated cases arise in practice (see, for example, Smith (1978)). Formally the same model can be given for potential and steady-state diffusion problems.

the assumption that u is sufficiently differentiable). The approximation is then made by dropping the two error terms and defining quantities $v_{i,j}$ for $0 \leqslant i, j \leqslant m$ to satisfy

$$\frac{v_{i+1,j} - 2v_{i,j} + v_{i-1,j}}{h^2} + \frac{v_{i,j+1} - 2v_{i,j} + v_{i,j-1}}{h^2} = 0$$

There is one such equation at each interior mesh point, where $v_{i,j}$ is an estimate of the unknown value $u(x_i, y_j)$.

The equation for $v_{i,j}$, $1 \leqslant i, j \leqslant m - 1$, can be simplified to

$$-v_{i,j-1} - v_{i-1,j} + 4v_{i,j} - v_{i+1,j} - v_{i,j+1} = 0$$

This is known as the 'five-point replacement formula' for Laplace's equation. It relates, for example, v_{11} to the values $v_{10}, v_{01}, v_{21}, v_{12}$ at the four neighbouring points (see figure 7.3). Taking the known boundary values to the right hand side, we have

$$4v_{11} - v_{21} - v_{12} = v_{10} + v_{01}$$

For $h = 1/4$ there are altogether nine such equations; these can be expressed in matrix form as

$$\mathbf{Ax} = \mathbf{b} \tag{7.15}$$

where

$$
\mathbf{A} = \left[
\begin{array}{ccc|ccc|ccc}
4 & -1 & 0 & -1 & 0 & 0 & & & \\
-1 & 4 & -1 & 0 & -1 & 0 & & \bigcirc & \\
0 & -1 & 4 & 0 & 0 & -1 & & & \\
\hline
-1 & 0 & 0 & 4 & -1 & 0 & -1 & 0 & 0 \\
0 & -1 & 0 & -1 & 4 & -1 & 0 & -1 & 0 \\
0 & 0 & -1 & 0 & -1 & 4 & 0 & 0 & -1 \\
\hline
 & & & -1 & 0 & 0 & 4 & -1 & 0 \\
 & \bigcirc & & 0 & -1 & 0 & -1 & 4 & -1 \\
 & & & 0 & 0 & -1 & 0 & -1 & 4 \\
\end{array}
\right]
\begin{array}{l}
\left.\rule{0pt}{2.5em}\right\} m-1 \\[2em]
\left.\rule{0pt}{5em}\right\} (m-1)^2
\end{array}
$$

$$
\mathbf{x} = \left[
\begin{array}{c}
v_{11} \\
v_{21} \\
v_{31} \\
\hline
v_{12} \\
v_{22} \\
v_{32} \\
\hline
v_{13} \\
v_{23} \\
v_{33}
\end{array}
\right]
\quad \text{and} \quad
\mathbf{b} = \left[
\begin{array}{c}
v_{10} + v_{01} \\
v_{20} \\
v_{30} \\
\hline
v_{02} \\
0 \\
v_{42} \\
\hline
v_{03} + v_{14} \\
v_{24} \\
v_{43} + v_{34}
\end{array}
\right]
$$

These equations could be represented and solved easily by our existing methods. However, to reduce the discretisation error we normally take a much smaller value for h. For $h = 1/m$ there are $n = (m-1)^2$ interior points, and we have an $n \times n$ system of linear equations to solve. The coefficient matrix in general has a similar structure to the matrix $\mathbf{A}$ in equation 7.15. There are $m-1$ square blocks on the diagonal, each of order $m-1$ and of tridiagonal form; there are blocks $-\mathbf{I}$ to each side, where $\mathbf{I}$ is the unit matrix of order $m-1$; all the other matrix elements are zero. This structure is called *block tridiagonal*. It is a band form, but many of the elements within the band are zero. The result is a very sparse matrix. For example, with $m = 20$ we have a matrix of order $(20-1)^2 = 361$ and a total of $130,321$ matrix elements. All but about 5×361 (that is $1\frac{1}{2}$ per cent) of these elements are zero.

Clearly, we would not store such matrices in full; there would even be a high redundancy if we stored them as band matrices. In fact we do not store them at all, since only the non-zero elements are of interest and these take only two distinct values and occur in a regular pattern. The elements can be computed when required.

We illustrate some important properties of the class of large, sparse matrices exemplified by $\mathbf{A}$ in equation 7.15. We have referred to a symmetric matrix $\mathbf{A}$ as *positive definite* if the scalar product $\mathbf{x}^T \mathbf{A} \mathbf{x}$ is positive for all non-zero vectors $\mathbf{x}$. As a simple example

$$[x\ y] \begin{bmatrix} 2 & 1 \\ 1 & 2 \end{bmatrix} \begin{bmatrix} x \\ y \end{bmatrix} = 2x^2 + 2xy + 2y^2$$
$$= (x+y)^2 + x^2 + y^2$$
$$> 0 \text{ unless } x = y = 0$$

Positive definiteness is often inferred on physical grounds; in a typical application the product $\mathbf{x}^T \mathbf{A} \mathbf{x}$ might represent potential energy or power dissipation, which is inherently positive. The matrix in equation 7.15 is positive definite.

An $n \times n$ matrix (a_{ij}) is said to be *diagonally dominant* if

$$|a_{ii}| \geqslant \sum_{\substack{j=1 \\ j \neq i}}^{n} |a_{ij}| \text{ for } 1 \leqslant i \leqslant n$$

with strict inequality for at least one i. The matrix in equation 7.15 is diagonally dominant, since every diagonal element is 4 and the sum of the absolute values of the other elements in any row is 2, 3 or 4.

A square matrix is said to be *reducible* if there exists a permutation of the rows and columns such that the matrix can be rearranged in the form

$$\begin{bmatrix} \mathbf{A} & \mathbf{B} \\ \mathbf{O} & \mathbf{C} \end{bmatrix}$$

where $\mathbf{A}$ and $\mathbf{C}$ are square blocks. Otherwise, the matrix is said to be *irreducible* (this means that no subset of equations with this coefficient matrix can be solved separately from the other equations in the set). The matrix in equation 7.15 is irreducible.

Elimination or factorisation methods applied to the large, sparse matrices under consideration would change many of the zero elements within the band to non-zero elements; worse, these would have complicated non-zero values which would have to be stored. To avoid this problem we seek methods which do not alter any of the matrix elements but preserve the sparseness of the original matrix. (A good up-to-date account of methods for solving large, sparse systems, particularly with positive definite coefficient matrices, is given in the book by George and Liu (1981).)

Iterative Methods

Iterative methods for solving linear systems generate a sequence of vectors $\left\{ \mathbf{x}^{(k)} \right\}$, $k = 0, 1, \ldots$, often starting with $\mathbf{x}^{(0)} = \mathbf{0}$; under suitable conditions the sequence converges to the solution $\mathbf{x}$ of $\mathbf{Ax} = \mathbf{b}$.

We consider the system of equations 7.1. If $a_{ii} \neq 0$ for $1 \leqslant i \leqslant n$, the equations can be rewritten as

$$x_i = \frac{1}{a_{ii}} \left(b_i - \sum_{\substack{j=1 \\ j \neq i}}^{n} a_{ij} x_j \right), \quad i = 1, \ldots, n$$

(the ith equation is 'solved' for x_i). Now if $\mathbf{x}^{(k)} = (x_1^{(k)}, \ldots, x_n^{(k)})^{\mathrm{T}}$ is an approximation to the solution $\mathbf{x}$, we can define an iterative step by

$$x_i^{(k+1)} = \frac{1}{a_{ii}} \left(b_i - \sum_{\substack{j=1 \\ j \neq i}}^{n} a_{ij} x_j^{(k)} \right), \quad i = 1, \ldots, n$$

This is the *Jacobi method*. Alternatively, we can define

$$x_i^{(k+1)} = \frac{1}{a_{ii}} \left(b_i - \sum_{j=1}^{i-1} a_{ij} x_j^{(k+1)} - \sum_{j=i+1}^{n} a_{ij} x_j^{(k)} \right), \quad i = 1, \ldots, n \qquad (7.16)$$

This is the *Gauss–Seidel method*.

In the Gauss–Seidel method the most up-to-date values known for all the components are used at each stage to compute the next estimate of x_i (in programming terms this means that only one vector needs to be kept for $\mathbf{x}$ and this is updated component-by-component in cyclic order). The Gauss–Seidel method often, but not always, converges faster than the Jacobi method.

As with all iterative methods, the question must be asked whether convergence will be achieved. It can be shown that if the matrix $\mathbf{A}$ is diagonally dominant and irreducible, then both the Jacobi and Gauss–Seidel methods converge to the solution of $\mathbf{Ax} = \mathbf{b}$ from any starting point. Furthermore, if $\mathbf{A}$ is positive definite, the Gauss–Seidel method converges. The matrix in equation 7.15 qualifies on both counts.

We shall not analyse the convergence of these methods but merely remark that it is useful to define an *iteration matrix* $\mathbf{M}$ relating successive vectors in the sequence†; thus

$$\mathbf{x}^{(k+1)} = \mathbf{M}\mathbf{x}^{(k)} + \mathbf{c}$$

where $\mathbf{c}$ is a constant vector. If the coefficient matrix is split into three parts

$$\mathbf{A} = \mathbf{L} + \mathbf{D} + \mathbf{U}$$

where $\mathbf{L}$ and $\mathbf{U}$ are the (strict) lower and upper triangular parts and $\mathbf{D}$ is the diagonal, then $(\mathbf{L} + \mathbf{D} + \mathbf{U})\mathbf{x} = \mathbf{b}$, and it is easy to show that the iteration matrices are

$$-\mathbf{D}^{-1}\,(\mathbf{L} + \mathbf{U}) \quad \text{(Jacobi method)}$$
$$-(\mathbf{L} + \mathbf{D})^{-1}\,\mathbf{U} \quad \text{(Gauss–Seidel method)}$$

From $\mathbf{x}^{(k+1)} - \mathbf{x} = \mathbf{M}(\mathbf{x}^{(k)} - \mathbf{x})$, it follows that a sufficient condition for convergence is

$$\|\mathbf{M}\| < 1$$

in any subordinate norm.

The rate of convergence of the basic iterative methods is usually very slow, particularly for large systems. For practical use it is essential to accelerate the rate of convergence. This may be done by introducing a *relaxation factor* ω. If we write the Gauss–Seidel iteration in equation 7.16 as

$$x_i^{(k+1)} = x_i^{(k)} + r_i^{(k)}$$

where

$$r_i^{(k)} = \frac{1}{a_{ii}} \left(b_i - \sum_{j=1}^{i-1} a_{ij} x_j^{(k+1)} - \sum_{j=i}^{n} a_{ij} x_j^{(k)} \right)$$

and regard $r_i^{(k)}$ as the 'Gauss–Seidel correction', we can modify the process by taking a multiple ω of this correction to form $x_i^{(k+1)}$. Thus

$$x_i^{(k+1)} = x_i^{(k)} + \omega r_i^{(k)}, \quad i = 1, \ldots, n$$

The factor ω is often taken to be a constant independent of k. For $\omega = 1$ this is still the Gauss–Seidel method. For $\omega \neq 1$ (usually $1 < \omega < 2$) the method is known as *successive overrelaxation (SOR)*. Assuming for simplicity that the matrix $\mathbf{A}$ is stored in an $n \times n$ array, we could express a complete cycle of SOR as

```
FOR i := 1 TO n DO
   x[i] := x[i] + omega * (b[i] - inprod (A, x, i, 1, n))
                                           / A[i,i]
```

The rate of convergence of SOR depends on the value used for the relaxation factor ω; the aim is to choose ω so as to accelerate and if possible optimise the rate of convergence. In some cases it is possible to calculate the optimal relaxa-

---

†Iterative methods for solving large, sparse systems are described and analysed by Varga (1962).

tion factor ω^* (see Varga (1962)). For Laplace's equation in a square region, with a square mesh of m subdivisions, it can be shown that

$$\omega^* = \frac{2}{1 + \sin(\pi/m)}$$

We give a table comparing the performance of the Gauss-Seidel method with that of SOR (with optimal relaxation factor) on the model problem in equation 7.15 for mesh sizes ranging from 1/10 to 1/30. The boundary values are all set to 1 so that the exact solution is known to be $x = (1, 1, \ldots, 1)^T$ (similar performance is obtained for other boundary conditions). The starting vector in all cases is $x^{(0)} = 0$, and the iteration is terminated when the maximum correction is less than 10^{-15}. The results on the Amdahl V/7 are

m	n	GAUSS-SEIDEL			SOR		
		Iterations	Max. error	Time (secs)	Iterations	Max. error	Time (secs)
10	81	328	10.0E−15	1.9	64	2.6E−15	0.4
15	196	723	26.1E−15	9.8	95	6.0E−15	1.4
20	361	1267	44.9E−15	31.8	126	10.7E−15	3.3
25	576	−	−	−	157	16.7E−15	6.4
30	841	−	−	−	188	23.0E−15	11.3

Much larger systems than these can arise in the solution of partial differential equations. Clearly, the Gauss-Seidel method is not a practical proposition and, if SOR is to be used, a good approximation to the optimal relaxation factor is required. In the next section we shall describe an attractive alternative method for solving large, sparse systems of this type which does not involve any such problem-dependent parameter.

7.6   THE CONJUGATE GRADIENT METHOD

The sum of squares of residuals in an $n \times n$ linear system,

$$r^T r = (b - Ax)^T (b - Ax)$$

is non-negative for all n-component vectors x. The minimum value of $r^T r$ is zero; this occurs when x is the exact solution of $Ax = b$, since then $r = b - Ax = 0$ and $r^T r = 0$. Instead of solving $Ax = b$ directly, we can consider the solution x as the point in n-dimensional space at which the function $r^T r$ attains its minimum. We can try to locate this point, starting at some initial approximation x_0 and taking a succession of steps in suitable directions which progressively reduce the value of $r^T r$†.

†Methods based on this idea are iterative; they have the same advantage for large, sparse systems as the iterative methods discussed earlier, in that they retain the original matrix A unmodified and so avoid storage problems.

If A is positive definite (therefore necessarily symmetric), as in the model problem for Laplace's equation and many other applications, the computation is simplified if we work with the function

$$\Phi(x) = r^T A^{-1} r$$

As A is positive definite so is A^{-1}, hence $\Phi(x)$ is non-negative; the minimum value, $\Phi(x) = 0$, is attained as before when $r = 0$. It is the corresponding vector x which we wish to determine.

Suppose we have an approximation x_k, in general with $\Phi(x_k) > 0$, and a search direction p_k (subscripts here will denote iteration numbers rather than components). We wish to minimise $\Phi(x)$ along the line p_k passing through x_k, in order to obtain the next approximation x_{k+1}. Any point along the line can be expressed in the form $x = x_k + \alpha p_k$, where α is a scalar parameter. Let the particular point on the line at which $\Phi(x)$ is minimum be denoted by $x^* = x_k + \alpha^* p_k$.

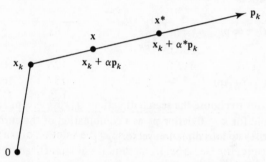

Figure 7.4 Unidirectional minimisation along p_k

The residual vector r at the point x in figure 7.4 can be expressed as

$$r = b - Ax = b - Ax^* + A(x^* - x)$$
$$= r^* + A(\alpha^* - \alpha)p_k$$

hence

$$A^{-1} r = A^{-1} r^* + (\alpha^* - \alpha)p_k$$

Taking the transpose of the first equation and postmultiplying by the second, we obtain

$$r^T A^{-1} r = r^{*T} A^{-1} r^* + r^{*T} (\alpha^* - \alpha)p_k$$
$$+ p_k^T A^T (\alpha^* - \alpha)A^{-1} r^* + p_k^T A^T (\alpha^* - \alpha)^2 p_k$$

Hence with $A^T = A$ and $AA^{-1} = I$, we have

$$\Phi(x) = \Phi(x^*) + 2(\alpha^* - \alpha)p_k^T r^* + (\alpha^* - \alpha)^2 p_k^T A p_k$$

where we have used the identity $r^{*T} p_k = p_k^T r^*$. Since A is positive definite, the last term on the right is non-negative. It follows that $\Phi(x) \geqslant \Phi(x^*)$, so that x^* is the required minimum point along the line, provided that $(\alpha^* - \alpha)\mathbf{p}_k^T \mathbf{r}^* = 0$ for all α. This condition is satisfied if $\mathbf{p}_k^T \mathbf{r}^* = 0$.

For $\mathbf{r}^*$ we can write

$$\mathbf{r}^* = \mathbf{b} - \mathbf{A}\mathbf{x}^* = \mathbf{b} - \mathbf{A}(\mathbf{x}_k + \alpha^* \mathbf{p}_k)$$
$$= \mathbf{r}_k - \alpha^* \mathbf{A}\mathbf{p}_k$$

Hence the unidirectional minimum is attained when

$$\mathbf{p}_k^T (\mathbf{r}_k - \alpha^* \mathbf{A}\mathbf{p}_k) = 0$$

that is

$$\alpha^* = \frac{\mathbf{p}_k^T \mathbf{r}_k}{\mathbf{p}_k^T \mathbf{A}\mathbf{p}_k}$$

Since $\mathbf{r}_k$ and $\mathbf{p}_k$ are known, this determines the scalar α^*. Thus we take as the new approximation

$$\mathbf{x}_{k+1} = \mathbf{x}_k + \alpha_k \mathbf{p}_k, \quad \alpha_k = \frac{\mathbf{p}_k^T \mathbf{r}_k}{\mathbf{p}_k^T \mathbf{A}\mathbf{p}_k} \tag{7.17}$$

We also note that

$$\mathbf{r}_{k+1} = \mathbf{r}_k - \alpha_k \mathbf{A}\mathbf{p}_k \tag{7.18}$$

It now remains to choose the search directions $\mathbf{p}_0, \mathbf{p}_1, \ldots, \mathbf{p}_k, \ldots$. We shall take $\mathbf{p}_0 = \mathbf{r}_0$, and for $k > 0$ define $\mathbf{p}_k$ as a combination of the current residual vector $\mathbf{r}_k$ and the previous direction vector $\mathbf{p}_{k-1}$

$$\mathbf{p}_k = \mathbf{r}_k + \beta_{k-1} \mathbf{p}_{k-1} \tag{7.19}$$

As explained below, the scalar parameter β_{k-1} is to be determined so that $\mathbf{p}_k$ and $\mathbf{p}_{k-1}$ are *conjugate* with respect to the matrix $\mathbf{A}$, that is, $\mathbf{p}_k^T \mathbf{A}\mathbf{p}_{k-1} = 0$. Thus

$$(\mathbf{r}_k^T + \beta_{k-1} \mathbf{p}_{k-1}^T) \mathbf{A}\mathbf{p}_{k-1} = 0$$

giving

$$\beta_{k-1} = - \frac{\mathbf{r}_k^T \mathbf{A}\mathbf{p}_{k-1}}{\mathbf{p}_{k-1} \mathbf{A}\mathbf{p}_{k-1}} \tag{7.20}$$

The algorithm specified by equations 7.19 and 7.17, with the given formulae for α_k and β_{k-1}, is called the *conjugate gradient method*.

A slightly modified formula will be used for α_k. We have, using equation 7.18

$$\mathbf{p}_{k-1}^T \mathbf{r}_k = \mathbf{p}_{k-1}^T (\mathbf{r}_{k-1} - \alpha_{k-1} \mathbf{A}\mathbf{p}_{k-1})$$
$$= 0 \text{ (by the formula for } \alpha_{k-1})$$

Then with equation 7.19 for $\mathbf{p}_k$

$$\mathbf{p}_k^T \mathbf{r}_k = (\mathbf{r}_k^T + \beta_{k-1} \mathbf{p}_{k-1}^T) \mathbf{r}_k$$

$$= \mathbf{r}_k^T \mathbf{r}_k + \beta_{k-1} \mathbf{p}_{k-1}^T \mathbf{r}_k$$

$$= \mathbf{r}_k^T \mathbf{r}_k \quad \text{(by the above result)}$$

Hence the formula for α_k in equation 7.17 may be rewritten as

$$\alpha_k = \frac{\mathbf{r}_k^T \mathbf{r}_k}{\mathbf{p}_k^T \mathbf{A} \mathbf{p}_k}$$

which will allow us to compute one less inner product in our library routine.

The reason for the particular choice of search directions in the conjugate gradient method is that, theoretically, this ensures convergence in a *finite* number of steps. It can be shown (see, for example, Fox (1964), chapter 8) that the n vectors $\mathbf{p}_0, \mathbf{p}_1, \ldots, \mathbf{p}_{n-1}$ form a linearly independent set, and the residual vector $\mathbf{r}_n$ is orthogonal to all of them, that is, $\mathbf{r}_n^T \mathbf{p}_k = 0$ for $0 \leqslant k \leqslant n-1$. Since the vector space is n-dimensional, it follows that $\mathbf{r}_n$ must be the zero vector, that is, $\mathbf{b} - \mathbf{A}\mathbf{x}_n = 0$. In theory, therefore, the search terminates with the exact solution of $\mathbf{A}\mathbf{x} = \mathbf{b}$ after n steps. This means that the conjugate gradient method is, strictly speaking, a direct method. In actual computation rounding errors upset this behaviour, and it is more useful to regard the method as iterative. For well-conditioned problems the method usually attains a sufficiently accurate solution in fewer than n steps, while for ill-conditioned problems more than n steps may be required.

The iterative scheme is summarised for later use as follows.

Initialisation

$\mathbf{x}_0 = \mathbf{0}$ (or a better approximation if one is known)

$\mathbf{r}_0 = \mathbf{b} - \mathbf{A}\mathbf{x}_0, \quad \mathbf{p}_0 = \mathbf{r}_0$

For $k = 0, 1, \ldots$

$$\mathbf{q}_k = \mathbf{A}\mathbf{p}_k, \quad \alpha_k = \frac{\mathbf{r}_k^T \mathbf{r}_k}{\mathbf{p}_k^T \mathbf{q}_k}$$

$$\mathbf{x}_{k+1} = \mathbf{x}_k + \alpha_k \mathbf{p}_k, \quad \mathbf{r}_{k+1} = \mathbf{r}_k - \alpha_k \mathbf{q}_k$$

$$\beta_k = - \frac{\mathbf{r}_{k+1}^T \mathbf{q}_k}{\mathbf{p}_k^T \mathbf{q}_k}, \quad \mathbf{p}_{k+1} = \mathbf{r}_{k+1} + \beta_k \mathbf{p}_k$$

Implementation of the Conjugate Gradient Method

Since in practice the conjugate gradient method is an iterative process, a terminating criterion is required. The aim is to continue the iteration until the growth of rounding error prevents any further improvement in the approximation $\mathbf{x}_k$. An indication of this is given by the residual vector $\mathbf{r}_k$. In the algorithm above, $\mathbf{r}_k$ is computed from $\mathbf{r}_{k-1}$, and because of accumulated rounding error this will be slightly different from the true residual $\mathbf{b} - \mathbf{A}\mathbf{x}_k$. Thus the squared norm of the difference,

$$\text{diffsq} = \| \mathbf{b} - \mathbf{A}\mathbf{x}_k - \mathbf{r}_k \|_2{}^2$$

is in general non-zero and increases with k. On the other hand the square of the computed residual, $\text{rr} = \| \mathbf{r}_k \|_2{}^2$, decreases fairly steadily as the solution is approached. When diffsq becomes about as large as rr, we have reached the round-off error level in $\mathbf{r}_k$ and the iteration should be terminated.

We shall use a criterion suggested in the article by T. Ginsburg in Wilkinson and Reinsch (1971): the iteration is terminated when

$$\text{rr} \leqslant \exp{(k/n)^2} \times \text{diffsq} \tag{7.21}$$

The rapidly increasing factor on the right allows early termination in well-conditioned problems (often with $k \ll n$) and also forces eventual termination in ill-conditioned problems.

The evaluation of diffsq, involving a product $\mathbf{A}\mathbf{x}_k$, is computationally expensive in comparison with the other operations of the algorithm; diffsq is therefore updated only every tenth step. The terminating condition 7.21 is tested after every step, using the current rr and the most recent value for diffsq.

Before we give a routine for the conjugate gradient method, we define some additional library data types. The reason for using the method is that it can deal with sparse systems much larger than those which can be treated by our earlier routines. We therefore define a new library constant, hibnd = 1000, and the following data types

```
(* To represent long vectors in sparse linear systems *)
longrange  = 1 .. hibnd ;  (* range of the array subscript *)
longvector = ARRAY [longrange] OF real ;
```

We also require a library utility function to compute inner products of long vectors.

```
FUNCTION longinprod (     n    : longrange ;
                      VAR x, y : longvector ) :  real ;
(* Computes the inner product x[j]*y[j], j = 1,...,n, where
   x and y are long vectors. Avoids underflow failure in
   individual products by using the library function mult *)

   VAR
      j   : longrange ;
      sum : real ;

   BEGIN
   sum := 0.0 ;
   FOR j := 1 TO n DO
      sum := sum + mult(x[j], y[j]) ;
   longinprod := sum
   END (* longinprod *)
```

We now give the library routine congrad for solving a system $\mathbf{A}\mathbf{x} = \mathbf{b}$ with positive definite coefficient matrix. The matrix is not supplied as a parameter in the usual way because it is not stored in an array; this is most important for economy of storage. Instead, **A** is defined by a rule to compute the product **Ay** for any given vector **y**. The user must provide the rule as a procedure in the calling program, and the name of this procedure is supplied as a parameter to congrad.

In studying the following code the reader is advised to refer to the algorithm stated earlier, which is followed fairly closely.

```
PROCEDURE congrad ( n : longrange ; (* order of the system     *)
            PROCEDURE aprod (n:longrange;
              VAR y,z:longvector) ; (* user-supplied procedure
                                      for matrix A            *)
            VAR b      : longvector ; (* right hand side vector *)
            VAR x.     : longvector ; (* solution vector        *)
            VAR count : integer ;    (* number of iterations   *)
            VAR eflag : integer   )  ;

(* Solves the nxn system of linear equations Ax = b by the
   conjugate gradient method. The coefficient matrix must be
   symmetric and positive definite; it is normally large and
   sparse. The matrix is not stored explicitly; instead, a
   procedure must be supplied in the calling program
   corresponding to the formal parameter aprod. The procedure
   should compute z = Ay for any n-component vector y.
   After a successful call of congrad, x contains an approximate
   solution of Ax = b, count records the number of iterations.
   The error number for eflag is:
     =1  The matrix is not positive definite  (fatal).
   Routines called from the library:
   longinprod, mult, errormessage      *)

VAR
    alpha, beta,
      denom, rr, diffsq : real ;
    i                   : longrange ;
    p, q, r             : longvector ;
    posdef              : boolean ;

BEGIN
(* Initialisation *)
posdef := true ;  (* switched to false if the matrix is
                     found to be not positive definite *)
FOR i := 1 TO n DO
    BEGIN
    x[i]  := 0.0 ;  r[i] := b[i] ;  p[i] := r[i]
    END ;
rr := longinprod (n, r, r) ;  diffsq := 0.0 ;

count := 0 ;
REPEAT  (* iteration loop *)
    aprod (n, p, q) ;   (* compute q = Ap *)
    denom := longinprod (n, p, q) ;
    IF denom <= 0.0 THEN
       posdef := false  (* matrix is not positive definite *)
    ELSE
       BEGIN  (* compute new vectors x, r and p *)
       count := count+1 ;
       alpha := rr/denom ;
       FOR i := 1 TO n DO
          BEGIN
          x[i]  := x[i] + mult(alpha, p[i]) ;
          r[i]  := r[i] - mult(alpha, q[i])
          END ;
       beta := - longinprod (n, r, q) / denom ;
       FOR i := 1 TO n DO
          p[i]  := r[i] + mult(beta, p[i]) ;
```

```
            rr := longinprod (n, r, r) ;
            IF count MOD 10 = 0 THEN
               BEGIN  (* compute new diffsq *)
               aprod (n, x, q) ;    (* q := Ax *)
               FOR i := 1 TO n DO
                  q[i] := b[i] - q[i] - r[i] ;
               diffsq := longinprod (n, q, q)
               END
         END
   UNTIL (rr <= exp(sqr(count/n))*diffsq)
      OR NOT posdef ;

   IF posdef THEN
      eflag := 0  (* successful call *)
   ELSE
      errormessage ('congrad ', 1,
         'matrix not positive definite  ', fatal, eflag)

   END (* congrad *)
```

Notes

(i)   The calculation of a product of the form **Ay**, that is, a call of aprod, is the most expensive part of the computation. There is only one such product, $\mathbf{Ap}_k$, in a normal iterative step.

(ii)   The scalar product $\mathbf{p}_k^T \mathbf{Ap}_k$ occurs as a denominator denom in the routine; if this is zero or negative, then **A** is not positive definite and the computation is abandoned with an error message.

We shall not give a full example program for congrad but shall merely show how the procedure aprod could be written for the model problem of equation 7.15 (the five-point replacement method for Laplace's equation in a square). procedure aprod must be declared in the user program; the program should also read in the number m of subdivisions, compute the order $n = (m - 1)^2$, and set up the right hand side vector b for the appropriate boundary values.

```
PROCEDURE aprod (     n    : longrange ;
                  VAR y, z : longvector ) ;
(* Computes the product z = Ay for a given n-component vector y,
   where A is the nxn matrix for Laplace's equation in a square
   using the five-point replacement formula *)

   VAR
      i, k, l : longrange ;

   BEGIN
   l := round(sqrt(n)) ;
   z[1] := 4*y[1] - y[2] - y[l+1] ;
   z[n] := 4*y[n] - y[n-1] - y[n-1] ;
   k := n-1 ;
   FOR i := 2 TO l DO
      BEGIN
      z[i] := 4*y[i] - y[i-1] - y[i+1] - y[i+1] ;
      z[k] := 4*y[k] - y[k-1] - y[k+1] - y[k-1] ;
      k := k-1
      END ;
   FOR i := l+1 TO n-l DO
      z[i] := 4*y[i] - y[i-1] - y[i+1] - y[i-1] - y[i+1] ;
   (* Correct the values at edge points *)
   k := l ;
```

```
FOR i := 1 TO l-1 DO
   BEGIN
   z[k] := z[k] + y[k+1] ;
   z[k+1] := z[k+1] + y[k] ;
   k := k+1
   END
END (* aprod *)
```

We give a table showing the performance of congrad on the model problem of equation 7.15 with boundary values set to 1 and mesh sizes ranging from 1/10 to 1/30.

m	n	Iterations	Maximum error	Time (secs)
10	81	13	1.5E−15	0.2
15	196	28	5.9E−15	0.9
20	361	46	13.6E−15	2.8
25	576	59	15.6E−15	5.7
30	841	72	12.2E−15	10.2

This should be compared with the table of results for successive overrelaxation at the end of section 7.5. For a system with positive definite coefficient matrix, the conjugate gradient method achieves a performance comparable to SOR without the need for the (problem-dependent) optimal relaxation factor.

The conjugate gradient method is widely used to solve the very large systems of linear equations which arise from the discretisation of differential equations. It is normal practice to *precondition* the matrix. For example, the system $Ax = b$ can be transformed to

$$L^{-1}A(L^{T})^{-1}y = c$$

where LL^{T} is an approximate factorisation of A, $y = L^{T}x$ and $c = L^{-1}b$. If the matrix L has the same sparsity pattern as the lower triangular part of A, it is found that the computational effort required to solve the transformed system by the conjugate gradient method is considerably less than for the original system (see references in the article by J. K. Reid in Jacobs (1978)).

EXERCISES

7.1   (i) Show, using the matrix multiplication rule, that if the matrix product AB is defined, then $(AB)^{T} = B^{T}A^{T}$.

(ii)   Show, using the definition of inverse, that if A and B are non-singular matrices of the same order, then $(AB)^{-1} = B^{-1}A^{-1}$.

(iii)   Show that if a square matrix A is non-singular then $(A^{-1})^{T} = (A^{T})^{-1}$. Deduce that the inverse of a non-singular symmetric matrix is symmetric.

7.2   Reduce the following matrix to upper triangular form by Gaussian elimination

$$\begin{bmatrix} 1 & c & c^2 \\ c & 1 & c \\ c^2 & c & 1 \end{bmatrix}$$

Can a symmetric matrix be singular? Under what circumstances is this matrix singular? Check the conditions of singularity.

7.3   In *Gauss–Jordan reduction* a square matrix is reduced directly to diagonal form by eliminations above the pivot as well as below; thus, back substitution is not required. Show that in solving an $n \times n$ system by this method the number of multiplicative (and additive) operations is asymptotically $\frac{1}{2}n^3$, that is, 50 per cent more than in Gaussian elimination. Might row interchanges be necessary and, if so, how could they be used? What happens if the matrix is singular?

7.4   A set of n- component vectors $\{u_1, \ldots, u_k\}$ is said to be *linearly dependent* if there exist scalars $\lambda_1, \ldots, \lambda_k$, not all zero, such that $\lambda_1 u_1 + \ldots + \lambda_k u_k = 0$. Show that if the columns of an $n \times n$ matrix **A** are linearly dependent, then the system of equations $Ax = 0$ has at least one non-zero solution **x**. (*Hint:* consider **A** as an array of n column vectors and use the definition above.) Show further that there is an infinity of solutions.

Describe carefully how one such solution could be found by a form of Gaussian elimination and back substitution. Illustrate for the system

$$2x_1 + x_2 + 3x_3 = 0$$
$$x_1 - x_2 + x_3 = 0$$
$$x_1 + 2x_2 + 2x_3 = 0$$

7.5   If an upper triangular matrix **U** is obtained from a square matrix **A** by Gaussian elimination with partial pivoting, then the determinant of **A** is equal to the product of the diagonal elements of **U** multiplied by $(-1)^s$, where s is the number of row interchanges in pivoting. Modify procedure lufac to return an integer parameter with the value +1 if the number s is even, -1 if s is odd. Hence write a program which uses the output from lufac to compute the determinant of an $n \times n$ matrix **A** (the triangular factors can be overwritten on **A** but the rows must be accessed through the vector row). As a precaution against overflow/underflow the required product should be accumulated in two parts, a normalised fraction and an exponent (you may find procedure float from example 1.5 useful).

Test the program on some low-order matrices, then compute the determinant of the 7×7 matrix in example 7.5.

7.6   A positive definite matrix **A** can be factorised in the form $A = LL^T$, where **L** is a lower triangular matrix with positive diagonal elements. Show, by expanding the product LL^T, that the elements of **L** can be determined from

$$l_{jj} = \left(a_{jj} - \sum_{k=1}^{j-1} l_{jk}{}^2\right)^{1/2}$$

$$l_{ij} = \left(a_{ij} - \sum_{k=1}^{j-1} l_{ik}l_{jk}\right)/l_{jj}, \quad i > j$$

Write a procedure cholfac to compute **L** for a positive definite matrix (scaling and pivoting should not be used). Consider what error conditions can arise, and provide suitable actions in your procedure.

7.7   Execute program testlineq (section 7.3) on the data of example 7.5. Explain any differences in the approximate and accurate solutions obtained from those in the example. Assuming that the condition number of the coefficient matrix is about 10^8, estimate roughly the precision of your computer (see section 7.4).

7.8   Consider carefully the requirements of the substitution routine lusolv if the coefficient matrix is factorised by procedure cholfac (exercise 7.6) instead of procedure lufac. Make copies of lusolv and linsolv with any necessary modifications. Hence write a program on the lines of testlineq to solve a system **Ax = b** when **A** is positive definite. Test on the data of example 7.5.

7.9   Theorem 7.1 suggests a method of computing the inverse of a non-singular matrix **A**. The objective is to find a matrix **X** satisfying **AX = I**. Take each column x_j of **X** and the corresponding column e_j of **I**, and solve the system $Ax_j = e_j$; repeat for $j = 1, 2, \ldots, n$. Write a procedure which calls lufac and linsolv to compute the accurate inverse of an $n \times n$ matrix by this method. Test on some low-order matrices, then compute the inverse of the Hilbert matrix H_6 (section 7.4). It is known that the elements of a Hilbert inverse are all integers; explain why the elements of your 'accurate' computed inverse are not.

7.10   Verify that the maximum vector norm satisfies the postulates 7.11 for a vector norm. Show that the maximum row-sum matrix norm has properties analogous to those in 7.11, and given an example of the triangle inequality for 2×2 matrices.

7.11   (i) Show that if **x** is the exact solution of a system **Ax = b** and $\bar{x}$ is an approximate solution with residual vector **r**, then

$$\frac{\|\bar{x} - x\|}{\|x\|} \leqslant \text{cond}(\mathbf{A}) \times \frac{\|r\|}{\|b\|}$$

(ii)   Show that every subordinate norm of a unit matrix is equal to 1. Hence prove that a lower bound on the condition number of a matrix in any subordinate norm is 1.

7.12 Determine values of x and y that best satisfy in the least-squares sense the system of equations

$$x - 2y = -1.04, \quad 3x - y = 1.96$$
$$2x + 3y = 5.07, \quad x + 4y = 4.93$$

Calculate the euclidean norm of the residual vector.

7.13 The non-zero elements of a tridiagonal matrix may be stored compactly in three vectors. Devise an efficient algorithm which uses Gaussian elimination and back substitution to solve a system $\mathbf{Ax} = \mathbf{b}$, where $\mathbf{A}$ is an $n \times n$ tridiagonal matrix stored in this way. Assume that $\mathbf{A}$ is positive definite so no pivoting is required. Show that the solution can be obtained in a total of $8n - 7$ arithmetic operations.

7.14 Show that if $\mathbf{B}$ is non-singular then $\mathbf{B}^T\mathbf{B}$ is symmetric and positive definite (see the definition in section 7.5). Hence construct some 2×2 positive definite matrices.

Prove that if a matrix $\mathbf{A}$ is positive definite then so is $\mathbf{A}^{-1}$. You may assume that $\mathbf{A}$ is non-singular, and note first from exercise 7.1(iii) that, as $\mathbf{A}$ is symmetric, so is $\mathbf{A}^{-1}$. Next establish the lemma: for any $\mathbf{x} \neq \mathbf{0}$ there is a $\mathbf{y} \neq \mathbf{0}$ such that $\mathbf{Ay} = \mathbf{x}$. Use this to prove the positive definiteness of $\mathbf{A}^{-1}$.

7.15 Write a program that calls procedure congrad to solve Laplace's equation in a unit square with the five-point replacement formula (section 7.5). Remember that a procedure corresponding to aprod must be declared in the main program, and the right hand side vector b must be set up for the appropriate boundary values. Test with the boundary values all set to 1; then solve the problem with the following boundary conditions

$$u(x, 0) = x^2 \quad u(0, y) = y$$
$$u(x, 1) = 1 \quad u(1, y) = 1$$

with 4, 8 and 16 subdivisions. In each case print out the estimate of $u(0.5, 0.5)$. Why are the three values different?

7.16 Write a procedure to solve a system $\mathbf{Ax} = \mathbf{b}$ by the method of successive overrelaxation. The coefficient matrix should not be stored in an array. The procedure should have a similar parameter list to congrad — with, however, aprod replaced by a function parameter to compute the inner product required in the SOR formula. There should also be a parameter for the relaxation factor, which must be supplied by the user. Use the library function tolmin to test for convergence, and also, as a precaution, impose a cut-off on the maximum number of iterations.

Use the procedure to solve Laplace's equation in a unit square with the five-point replacement formula. Test with the boundary values all set to 1, then solve the boundary value problem of exercise 7.15.

8 Numerical Integration: Fixed-point Rules

And thin partitions do their bounds divide.

John Dryden, *Absalom and Achitophel*

A recurring problem in science and engineering is to integrate a given function of one real variable. Such a function may be defined in one of two ways: by the approximate numerical values at a set of discrete points (*numerically defined function*) or by a formula (*analytically defined function*). In this and the following chapter we shall be concerned mainly with developing library codes to integrate analytically defined functions. Our aim will be to write routines which can compute the value of a definite integral to a specified accuracy without requiring any specialised knowledge from the user.

A *definite integral*, $\int_a^b f(x)\,dx$, is defined as the area under the curve $y = f(x)$ bounded by the x axis and the lines $x = a$ and $x = b$ (the mathematical definition is in terms of a limit — see, for example, Courant and John (1965)). We know from calculus that the value of a definite integral is given by

$$I = \int_a^b f(x)\,dx$$
$$= F(b) - F(a)$$

where $F(x)$ is any function satisfying $F'(x) \equiv f(x)$. In practice it is rarely possible to find such an 'antiderivative' even if a formula is known for $f(x)$. We must then resort to numerical methods.

In numerical integration we attempt to estimate the area under the curve by evaluating the *integrand* $f(x)$ at a set of distinct points $\{x_0, x_1, \ldots, x_n\}$, where $x_i \in [a, b]$ for $0 \leqslant i \leqslant n$. The weighted sum

$$I_n = \sum_{i=0}^{n} w_i f(x_i) \tag{8.1}$$

with a suitable choice of the weights w_i, is taken as an estimate of the integral I. The formula in equation 8.1 is called a *numerical integration* or *quadrature rule*. The estimate obtained from a rule of the form 8.1 is in general not exact, and we therefore examine the *truncation* or *discretisation error*

$$E_n = I - I_n$$

189

We should bear in mind that most of the computation involved in equation 8.1 is performed in evaluating the function $f(x)$. We therefore seek formulae which, as far as possible, reduce the number of evaluations of $f(x)$ for a given accuracy.

8.1   THE TRAPEZOIDAL RULE

One of the simplest ways to estimate an integral $I = \int_a^b f(x)\,dx$ is to employ linear interpolation, that is, to approximate the curve $y = f(x)$ by a straight line $y = p_1(x)$ passing through the points $(a, f(a))$ and $(b, f(b))$ and then to compute the area under the line. This is illustrated in figure 8.1.

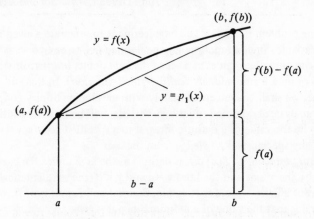

Figure 8.1   The trapezoidal rule

From the figure we see that

$$\int_a^b f(x)\,dx \approx \int_a^b p_1(x)\,dx$$

$$= (b - a)\left[f(a) + \frac{f(b) - f(a)}{2}\right]$$

$$= \frac{(b - a)}{2}\,[f(a) + f(b)] \tag{8.2}$$

This is known as the *trapezoidal rule*.

Since the integrand $f(x)$ is in general non-linear, the trapezoidal rule will not produce accurate estimates unless the interval $[a, b]$ is fairly small. For this reason we usually subdivide $[a, b]$ into a number of subintervals of equal *stepwidth* and apply the trapezoidal rule over each subinterval. This is shown in figure 8.2. A more accurate estimate of the integral is normally obtained by taking a large number of subintervals (that is, a small stepwidth).

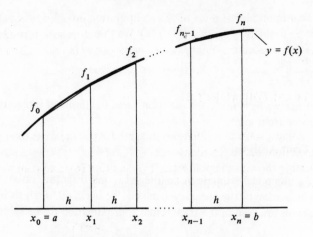

Figure 8.2   The composite trapezoidal rule

We define

$$x_0 = a, \; x_i = x_0 + ih \; \text{ for } \; i = 0, 1, \ldots, n$$
$$x_n = b$$

where the stepwidth h is given by

$$h = (b - a)/n$$

and n is the number of subintervals. Applying the trapezoidal rule in equation 8.2 to each subinterval, we find that

$$I \approx \sum_{i=1}^{n} \frac{h}{2} \; [f(x_{i-1}) + f(x_i)]$$

$$= h \left(\tfrac{1}{2} f_0 + f_1 + f_2 + \ldots + f_{n-1} + \tfrac{1}{2} f_n \right) \tag{8.3}$$

The abbreviation f_i is used for $f(x_i)$. The formula in equation 8.3, which will be denoted by $T(h)$, is called the *composite trapezoidal rule*.

To program the composite trapezoidal rule we first declare a Pascal function

```
FUNCTION f ( x : real ) :   real
```

to define the integrand. We may then use the following procedure.

```
PROCEDURE trapint (
           n      : integer ; (* number of subintervals    *)
      VAR ivalue : real      (* computed value of integral *)
                  ) ;
     VAR
        h, sum : real   ;
        i      : integer ;
```

```
BEGIN
h := (b-a)/n ;           (* stepwidth *)
sum := (f(a) + f(b))/2 ;
FOR i := 1 TO n-1 DO
   sum := sum + f(a+i*h) ;
ivalue := h * sum ;   (* estimate of integral *)
END (* trapint *)
```

For convenience we have made the function f and the limits of integration a, b global to procedure trapint.

Truncation Error Analysis

To compute a numerical estimate of an integral is unsatisfactory unless we can also give an indication of the accuracy attained. We begin our analysis by considering the error involved in the trapezoidal approximation to the subintegral

$$I^{(i)} = \int_{x_{i-1}}^{x_i} f(x)\,dx$$

The integrand $f(x)$ is approximated over $[x_{i-1}, x_i]$ by a linear polynomial $p_1^{(i)}(x)$ passing through the points (x_{i-1}, f_{i-1}) and (x_i, f_i). The following integral is then computed

$$T^{(i)} = \int_{x_{i-1}}^{x_i} p_1^{(i)}(x)\,dx = \frac{h}{2}(f_{i-1} + f_i)$$

An expression for the truncation error $I^{(i)} - T^{(i)}$ is given by

$$E^{(i)} = \int_{x_{i-1}}^{x_i} f(x)\,dx - \int_{x_{i-1}}^{x_i} p_1^{(i)}(x)\,dx$$

$$= \int_{x_{i-1}}^{x_i} (f(x) - p_1^{(i)}(x))\,dx \qquad (8.4)$$

To proceed we assume that $f(x)$ has the properties required for theorem 3.3 (essentially that f is twice differentiable in (x_{i-1}, x_i)). We may then write for the interpolation error

$$f(x) - p_1^{(i)}(x) = \frac{(x - x_{i-1})(x - x_i)}{2} f''(\xi_i), \quad \xi_i \in (x_{i-1}, x_i)$$

Equation 8.4 may now be rewritten as

$$E^{(i)} = \tfrac{1}{2} \int_{x_{i-1}}^{x_i} (x - x_{i-1})(x - x_i) f''(\xi_i)\,dx$$

The point ξ_i depends on x, and so $f''(\xi_i)$ cannot be taken directly outside the integral. However, we note that the factor $(x - x_{i-1})(x - x_i)$ does not change sign over (x_{i-1}, x_i). With the further assumption that f'' is continuous on $[x_{i-1}, x_i]$, we may use the *mean value theorem for integrals* (see, for example,

Phillips and Taylor (1973), chapter 2). This allows us to write the error expression as

$$E^{(i)} = \tfrac{1}{2} f''(\bar{\xi}_i) \int_{x_{i-1}}^{x_i} (x - x_{i-1})(x - x_i)\, dx$$

$$= -\frac{h^3}{12} f''(\bar{\xi}_i)$$

where $\bar{\xi}_i$ is a fixed (unknown) point in (x_{i-1}, x_i). This result can be used to obtain an expression for the truncation error of the composite trapezoidal rule.

Theorem 8.1

Consider the integral $I = \int_a^b f(x)\, dx$, where f is assumed to be twice continuously differentiable on $[a, b]$. The truncation error of the composite trapezoidal rule $T(h)$ with stepwidth h (equation 8.3) is given by

$$I - T(h) = -\frac{h^2}{12} (b - a) f''(\xi), \quad \xi \in (a, b)$$

Proof

The truncation error of the composite trapezoidal rule is the sum of the truncation errors for the n subintegrals. Thus

$$E_T = I - T(h) = \sum_{i=1}^{n} E^{(i)}$$

$$= -\frac{h^3}{12} \sum_{i=1}^{n} f''(\bar{\xi}_i), \quad \bar{\xi}_i \in (x_{i-1}, x_i)$$

Since f'' is continuous on $[a, b]$, there exists a maximum value $M = \max_{x \in [a,b]} f''(x)$ and a minimum value $m = \min_{x \in [a,b]} f''(x)$. Thus

$$nm \leqslant \sum_{i=1}^{n} f''(\bar{\xi}_i) \leqslant nM$$

It then follows by the intermediate value theorem (theorem 3.1) that there exists a point $\xi \in (a, b)$ such that

$$\sum_{i=1}^{n} f''(\bar{\xi}_i) = nf''(\xi)$$

Hence we can write the total error as

$$E_T = -\frac{h^3}{12} nf''(\xi) = -\frac{h^2}{12}(b - a) f''(\xi), \quad \xi \in (a, b)$$

(where we have used $nh = b - a$). $\square$

It follows immediately from theorem 8.1 that, under the stated conditions, the composite trapezoidal rule is a *second-order approximation process*, that is, $E_T = 0(h^2)$. However, since the point ξ in the statement of the theorem is unknown, we still do not have a computable expression for the error. One way of resolving this problem is to compute a *bound* on the truncation error. Suppose we know a bound on the second derivative

$$M_2 \geqslant \max_{x \in (a,b)} |f''(x)|$$

then from theorem 8.1 we have

$$|I - T(h)| \leqslant \frac{h^2}{12} |b - a| M_2 \tag{8.5}$$

For the integral

$$I = \int_1^2 (e^{-x}/x) \, dx$$

we have

$$f(x) = e^{-x}/x, \quad f'(x) = -\left(\frac{1}{x} + \frac{1}{x^2}\right) e^{-x}, \quad f''(x) = \left(\frac{1}{x} + \frac{2}{x^2} + \frac{2}{x^3}\right) e^{-x}$$

The second derivative attains its maximum value on $[1, 2]$ at $x = 1$; hence we can take

$$M_2 = f''(1) = 5e^{-1} \approx 1.84$$

Example 8.1

Write a program to estimate the integral $\int_1^2 (e^{-x}/x) \, dx$ using procedure **trapint** with n = 1, 2, 4, 8, 16, . . ., 256. Compare the actual error with the error bound given by formula 8.5, taking $M_2 = 1.84$ (the exact value of the integral is 0.170483423687. . .).

```
PROGRAM trapexample ( output ) ;

CONST
    a = 1.0 ;  b = 2.0 ;        (* interval of integration        *)
    maxsubints =   256 ;        (* maximum number of subintervals *)
    iexact = 0.170483423687 ;   (* exact value of integral        *)
    seconddervbnd = 1.84 ;      (* bound on second derivative
                                   in the interval (1, 2)          *)

VAR
    h, error, errbnd, ivalue : real ;
    n : integer ;

FUNCTION f ( x : real ) :  real ;
  BEGIN
  f := exp(-x)/x
  END ;
```

```
<PROCEDURE trapint> ;

BEGIN
writeln ('    h           n            ivalue
         'error    error bound') ;
n := 1 ;
REPEAT
    trapint (n, ivalue) ;              (* estimate the integral  *)
    error := abs(iexact-ivalue) ;      (* compute absolute error *)
    h := (b-a)/n ;                     (* stepwidth used in trapint *)
    errbnd := abs(b-a) * sqr(h) * seconddervbnd/12 ;
    writeln (h:10, n:6, ivalue:18:12, '  ', error:10, '  ',
                                                   errbnd:10) ;
    n := 2*n    (* double the number of subintervals *)
UNTIL n > maxsubints
END .
```

The results are

h	n	ivalue	error	error bound
1.00E+00	1	0.217773541395	4.73E-02	1.53E-01
5.00E-01	2	0.183263490747	1.28E-02	3.83E-02
2.50E-01	4	0.173757553810	3.27E-03	9.58E-03
1.25E-01	8	0.171307407474	8.24E-04	2.40E-03
6.25E-02	16	0.170689769993	2.06E-04	5.99E-04
3.12E-02	32	0.170535032321	5.16E-05	1.50E-04
1.56E-02	64	0.170496327227	1.29E-05	3.74E-05
7.81E-03	128	0.170486649659	3.23E-06	9.36E-06
3.91E-03	256	0.170484230186	8.07E-07	2.34E-06

The $O(h^2)$ behaviour of the trapezoidal rule can be seen clearly from the numerical results: each time n is doubled, the error is reduced by a factor of approximately four. □

The formula 8.5 can also be used to find the stepwidth required to achieve a given accuracy. If an accuracy of t decimal places is required, then from chapter 4 we use an absolute error test

$$|I - T(h)| \leqslant 0.5 \times 10^{-t}$$

This condition is satisfied if

$$\frac{h^2}{12} |b - a| M_2 \leqslant 0.5 \times 10^{-t}$$

that is

$$h \leqslant \sqrt{\left(\frac{6 \times 10^{-t}}{|b - a| M_2} \right)}$$

For the integral in example 8.1 we have $|b - a| = 1$ and $M_2 = 1.84$; for $t = 3$ we find that $h \leqslant 5.7 \times 10^{-2}$. Thus a convenient stepwidth for an accuracy of 3 decimal places would be $h = 1/20$ (the actual error is then 1.32×10^{-4}).

Numerical Estimation of Truncation Error

In practice it is at best inconvenient and often impossible to obtain the second derivative of the integrand $f(x)$; even if we can differentiate $f(x)$, it may still be very difficult to obtain a realistic bound. We must look for another approach which avoids explicit differentiation of the integrand. Two methods are commonly used: the first is to calculate an approximate bound M_2 numerically; the second is to estimate the truncation error by comparing the trapezoidal approximations obtained using two different stepwidths (normally h and $h/2$).

In the first method we estimate the second derivative $f''(x)$ numerically. It can be shown by Taylor's series expansion (see example 3.11) that if f is sufficiently differentiable then

$$f''(x) = \frac{f(x+h) - 2f(x) + f(x-h)}{h^2} + 0(h^2)$$

This formula can be used to approximate f'' at each of the integration points in (a, b). An approximate bound on $|f''(x)|$ can be found by computing

$$\max_{1 \leqslant i \leqslant n-1} \frac{|f(x_{i+1}) - 2f(x_i) + f(x_{i-1})|}{h^2}$$

An approximate bound on the truncation error is then obtained from formula 8.5. This method of estimating the error is particularly useful when the integrand is defined by a set of numerical values (for example, experimental data) and the stepwidth h is predetermined by the data points. However, care should be taken if h is small, as the formula for estimating f'' becomes very inaccurate when h is close to the relative precision of the computer.

The second method requires us to compute two estimates of the integral using two different stepwidths, h and $h/2$. From theorem 8.1 we have

$$I - T(h) = -\frac{(b-a)}{12} h^2 f''(\xi_1), \quad \xi_1 \in (a, b)$$

and

$$I - T(h/2) = -\frac{(b-a)}{12} \left(\frac{h}{2}\right)^2 f''(\xi_2), \quad \xi_2 \in (a, b)$$

If f is sufficiently smooth (meaning, in this case, that $f''(x)$ changes sufficiently slowly over $[a, b]$), we can make the assumption that $f''(\xi_1) \approx f''(\xi_2)$ (we shall justify this assumption in chapter 9). We can then eliminate $f''(\xi)$ between the two equations to give

$$4 \left(I - T(h/2)\right) \approx I - T(h)$$

hence

$$I - T(h/2) \approx \frac{T(h/2) - T(h)}{3} \tag{8.6}$$

This is a computable approximation to the truncation error $I - T(h/2)$.

The reason for using stepwidths h and $h/2$ is that the integration points for $T(h)$ coincide with a subset of those for $T(h/2)$. This results in a saving in the number of evaluations of $f(x)$. In general

$$T(h) = h \left\{ \tfrac{1}{2} f(a) + f(a + h) + \ldots + f(a + (n - 1)h) + \tfrac{1}{2} f(b) \right\}$$

and

$$T(h/2) = \frac{h}{2} \left\{ \tfrac{1}{2} f(a) + f(a + h/2) + f(a + h) + \ldots + f(a + (n - 1)h) \right.$$
$$\left. + f(a + (n - 1/2)h) + \tfrac{1}{2} f(b) \right\}$$

Thus we can write

$$T(h/2) = \tfrac{1}{2} T(h) + \frac{h}{2} \sum_{i=1}^{n} f(a + (i - 1/2)h) \tag{8.7}$$

The 'new' integration points for $T(h/2)$ are equally spaced between the original points for $T(h)$. If we had used a procedure based on equation 8.7 in example 8.1 instead of procedure trapint, we could have reduced the number of function evaluations by a factor of approximately two.

In chapter 9 we shall develop these ideas and use the formula 8.6 to determine automatically the stepsize required for a given accuracy.

8.2 SIMPSON'S RULE

The results of example 8.1 show that the composite trapezoidal rule, being only $0(h^2)$, requires a small stepwidth and hence a large number of evaluations of $f(x)$ for quite modest accuracy. Since function evaluations are nearly always computationally expensive, we seek *higher order* integration rules which require fewer evaluations of $f(x)$ for comparable accuracy.

To derive the composite trapezoidal rule we approximated the integrand by straight line segments, as shown in figure 8.2.† If instead we approximate the integrand by a set of higher degree polynomials (for example quadratics), we might expect, provided $f(x)$ is sufficiently differentiable, to obtain a closer fit to the curve $y = f(x)$. Since the error is smaller, we can derive an integration rule which allows us to achieve the same accuracy for a larger stepwidth.

There are several families of integration rules based on the idea of approximating the integrand by polynomials of increasing degree. The rules we shall

†We say that the curve $y = f(x)$ is represented by a continuous, piecewise linear approximation.

consider here are called *closed Newton–Cotes formulae* and take the general
form

$$\int_{x_0}^{x_p} f(x)\,dx = \sum_{i=0}^{p} w_i f(x_i) + E_p \qquad (8.8)$$

The integration points are equally spaced, that is, $x_i = x_0 + ih$ for $i = 0, 1, \ldots, p$,
with $h = (x_p - x_0)/p$. The *weights* w_i in equation 8.8 are chosen so that the integration rule is exact, that is, the truncation error $E_p = 0$, when $f(x)$ is a polynomial of degree $\leqslant p$.

Formulating this for the case $p = 1$, we obtain a rule $\int_{x_0}^{x_1} f(x)\,dx \approx w_0 f_0 +$
$w_1 f_1$ which should be exact when $f(x)$ is a straight line, that is, $f(x) \equiv a_0 + a_1 x$.
Substituting into equation 8.8 with $E_p = 0$ and equating coefficients of a_0 and
a_1 on both sides of the equation, we find that $w_0 = w_1 = h/2$. Thus, the Newton–
Cotes formula for $p = 1$ is simply the trapezoidal rule of equation 8.2. We have
seen that the truncation error in this case is $-h^3 f''(\xi)/12$, $\xi \in (x_0, x_1)$, confirming that the rule is exact when $f(x)$ is a linear polynomial.

Example 8.2

Determine the weights for the Newton–Cotes formula in the case $p = 2$.

Without loss of generality we can take the origin as $x_0 = 0$. We then have the
rule

$$\int_0^{2h} f(x)\,dx \approx w_0 f(0) + w_1 f(h) + w_2 f(2h)$$

The weights w_0, w_1, w_2 are to be chosen so that the rule is exact when $f(x)$ is a
polynomial of degree 2, that is, $f(x) \equiv a_0 + a_1 x + a_2 x^2$. The exact value of the
integral is then

$$\int_0^{2h} (a_0 + a_1 x + a_2 x^2)\,dx = a_0\,2h + a_1\,2h^2 + a_2\,8h^3/3$$

and the right hand side of the rule gives

$$w_0\,a_0 + w_1\,(a_0 + a_1 h + a_2 h^2) + w_2\,(a_0 + 2a_1 h + 4a_2 h^2)$$

Since the two sides are required to be equal for all $a_0, a_1, a_2 \in R$, we can equate
coefficients of a_0, a_1 and a_2 to give

$$2h \quad\ = w_0 + w_1 + w_2$$
$$2h^2 \quad = w_1 h + 2w_2 h$$
$$8h^3/3 = w_1 h^2 + 4w_2 h^2$$

A little manipulation shows that

$$w_0 = h/3, \quad w_1 = 4h/3, \quad w_2 = h/3$$

giving the rule

$$\int_{x_0}^{x_2} f(x)\,dx = \frac{h}{3}\,[f(x_0) + 4f(x_1) + f(x_2)] + E_2 \quad \square \tag{8.9}$$

The integration rule in equation 8.9 is known as *Simpson's rule*. The truncation error can be derived in a way similar to that of the trapezoidal rule (see, for example, Phillips and Taylor (1973), chapter 6) and is given by

$$E_2 = -\frac{h^5}{90}f^{(4)}(\xi), \quad \xi \in (x_0, x_2)$$

This result is rather unexpected and is something of a bonus; the occurrence of the fourth derivative shows that Simpson's rule is exact even when $f(x)$ is a cubic polynomial.

Other Newton–Cotes formulae can be derived similarly. For $p = 3$ we obtain the *3/8 rule*

$$\int_{x_0}^{x_3} f(x)\,dx = \tfrac{3}{8}h\,[f(x_0) + 3f(x_1) + 3f(x_2) + f(x_3)] + E_3$$

where

$$E_3 = -\frac{3}{80}h^5 f^{(4)}(\xi), \quad \xi \in (x_0, x_3)$$

Other things being equal, this rule is less attractive than Simpson's rule because both rules are of the same order of accuracy but the 3/8 rule requires one extra evaluation of $f(x)$.

Composite Simpson's Rule

As in the case of the composite trapezoidal rule, for greater accuracy the whole range of integration $[a, b]$ may be partitioned into n subintervals of equal width so that

$$x_0 = a, \quad x_i = x_0 + ih \quad \text{for} \quad i = 0, 1, \ldots, n$$

$$x_n = b$$

where the stepwidth h is given by

$$h = (b - a)/n$$

Provided n is *even* we can write the integral over $[a, b]$ as

$$I = \int_a^b f(x)\,dx = \int_{x_0}^{x_2} f(x)\,dx + \int_{x_2}^{x_4} f(x)\,dx + \ldots + \int_{x_{n-2}}^{x_n} f(x)\,dx$$

which gives a total of $n/2$ subintegrals. We can then apply Simpson's rule from equation 8.9 to each subintegral to obtain

$$I \approx \frac{h}{3} [(f_0 + 4f_1 + f_2) + (f_2 + 4f_3 + f_4) + \ldots + (f_{n-2} + 4f_{n-1} + f_n)]$$

$$= \frac{h}{3} (f_0 + 4f_1 + 2f_2 + 4f_3 + 2f_4 + \ldots + 2f_{n-2} + 4f_{n-1} + f_n) \qquad (8.10)$$

The formula in equation 8.10, which will be denoted by $S(h)$, is called the *composite Simpson's rule*.

The truncation error of the composite Simpson's rule is obtained by summing the truncation errors of the $n/2$ subintegrals,

$$E_S = I - S(h)$$

$$= -\frac{h^5}{90} \sum_{i=1}^{n/2} f^{(4)}(\xi_i), \quad \xi_i \in (x_{2i-2}, x_{2i})$$

With the assumption that $f^{(4)}$ is continuous on $[a, b]$, we can simplify this expression as we did in the proof of theorem 8.1 to obtain

$$E_S = -\frac{h^5}{90} \frac{n}{2} f^{(4)}(\xi) = -\frac{h^4}{180} (b - a) f^{(4)}(\xi), \quad \xi \in (a, b)$$

We see that, provided f if four times continuously differentiable on $[a, b]$, the composite Simpson's rule is a *fourth-order process*, that is, $E_S = 0(h^4)$.

To compare the performance of the rule in equation 8.10 with that of the composite trapezoidal rule in equations 8.3, we again consider the integral

$$I = \int_1^2 (e^{-x}/x) \, dx$$

The fourth derivative of the integrand is

$$f^{(4)}(x) = \left(\frac{1}{x} + \frac{4}{x^2} + \frac{12}{x^3} + \frac{24}{x^4} + \frac{24}{x^5} \right) e^{-x}$$

and this is bounded on $(1, 2)$ by

$$M_4 = 65e^{-1} \approx 24$$

Thus the truncation error satisfies

$$|E_S| \leqslant \frac{h^4}{180} |b - a| M_4 \approx \frac{24}{180} h^4$$

The numerical results obtained using the composite Simpson's rule with $n = 2, 4, 8, \ldots, 128$ are

h	n	ivalue	error	error bound
5.00E-01	2	0.171760140531	1.28E-03	8.33E-03
2.50E-01	4	0.170588908164	1.05E-04	5.21E-04
1.25E-01	8	0.170490692028	7.27E-06	3.26E-05
6.25E-02	16	0.170483890833	4.67E-07	2.03E-06
3.12E-02	32	0.170483453098	2.94E-08	1.27E-07
1.56E-02	64	0.170483425529	1.85E-09	7.95E-09
7.81E-03	128	0.170483423803	1.23E-10	4.97E-10

These results should be compared with those of example 8.1 for the composite trapezoidal rule. The number of evaluations of $f(x)$ for any given value of n is the same in both cases. The $0(h^4)$ behaviour of Simpson's rule can be seen clearly: each time n is doubled, the error is reduced by a factor of approximately $2^4 = 16$. We also see that the bound on the truncation error, $24h^4/180$, is a fairly good indicator of the actual error and could be used to determine the stepwidth for a specified accuracy in advance of the calculation.

Procedure simpint

Our first library routine for numerical integration is based on Simpson's rule and is designed to approximate the value of an integral $\int_a^b f(x)\,dx$ when $f(x)$ is *numerically defined* (for example, as experimental data). The routine requires the user to supply the numerical values of the integrand at n + 1 equally spaced points as y[0], y[1], . . ., y[n] and also the value of the stepwidth h. If n is even the integral is computed using Simpson's rule. If n is odd and greater than or equal to 3, one application of the 3/8 rule is used, and Simpson's rule is used for the remaining subintervals if any.

```
PROCEDURE simpint ( n        : range ;    (* number of subintervals *)
                    h        : real  ;    (* stepwidth              *)
                    VAR y    : coeff ;    (* function values        *)
                    VAR ivalue: real      (* computed integral      *)
                             ) ;

(* This procedure estimates the integral of a function whose
   numerical values y[0], y[1], ..., y[n], n <= upbnd, are given
   at equally-spaced points with stepwidth h. Repeated Simpson's
   rule is used, with one application of the 3/8 rule if n is odd
   (if n=1 the trapezoidal rule is used). The computed value of
   the integral is returned in ivalue. There are no error numbers
   and no indicator of accuracy in simpint *)

VAR
    i   : range ;
    sum : real ;

FUNCTION simprule ( i : index ) : real ;
    BEGIN
    simprule := y[i] + 4*y[i+1] + y[i+2]
    END ;


BEGIN  (* body of simpint *)
IF n = 1 THEN
    ivalue := (h/2) * (y[0] + y[1])  (* trapezoidal rule *)
```

```
ELSE
   BEGIN
   IF odd(n) THEN
      BEGIN  (* start with 3/8 rule *)
      sum := (9/8) * (y[0] + 3*(y[1]+y[2]) + y[3]) ;
      i := 3
      END
   ELSE
      BEGIN  (* start with Simpson's rule *)
      sum := simprule(0) ;
      i := 2
      END ;
   (* Continue with Simpson's rule *)
   WHILE i < n DO
      BEGIN
      sum := sum + simprule(i) ;
      i := i + 2
      END ;
   ivalue := (h/3)*sum
   END
END (* simpint *)
```

Notes

It is usually the case when $f(x)$ is numerically defined that the function values are known only approximately. In such cases there is often no advantage to be gained from using a formula of higher order than Simpson's rule. The errors in the numerical values of the integrand often dominate the truncation error of the integration rule; hence for this routine we do not provide an estimate of the error. Care is required in interpreting the results of simpint; when $f(x)$ is defined by a formula, it is preferable to use one of the mathlib routines to be developed in chapter 9.

8.3   GAUSSIAN RULES

If we relax the condition of equal spacing for the integration points, we can attempt to choose the points as well as the weights in an optimal manner. In a rule of the form

$$\int_{-1}^{1} f(x)\, dx = \sum_{i=1}^{p} w_i f(x_i) + E_p^{(G)}, \quad p \geqslant 1 \tag{8.11}$$

we can choose the points $x_1, \ldots, x_p$ and the weights $w_1, \ldots, w_p$ to make the rule exact for all polynomials of degree $\leqslant 2p - 1$ (such polynomials are determined by at most $2p$ coefficients). Rules of the form of equation 8.11, in which the integration points are chosen in this way, are called *Gaussian rules*. We shall see later how to transform the range of integration to any finite interval $[a, b]$; for the moment we restrict ourselves to the interval $[-1, 1]$.

For $p = 1$ we have the *1-point Gaussian rule*

$$\int_{-1}^{1} f(x)\, dx = w_1 f(x_1) + E_1^{(G)}$$

which is to be exact when $f(x)$ is a linear polynomial, $f(x) \equiv a_0 + a_1 x$. Equating coefficients of a_0 and a_1, we find

$$w_1 = \int_{-1}^{1} 1 \times dx = 2$$

$$w_1 x_1 = \int_{-1}^{1} x \, dx = 0 \quad \text{(hence } x_1 = 0)$$

Thus the 1-point Gaussian rule is given by

$$\int_{-1}^{1} f(x) \, dx = 2f(0) + E_1^{(G)}$$

This is simply the mid-point rule of example 1.1 (chapter 1).

Example 8.3

Determine the weights and integration points for the 2-point Gaussian rule.
  We have

$$\int_{-1}^{1} f(x) \, dx = w_1 f(x_1) + w_2 f(x_2) + E_2^{(G)}$$

The weights w_1, w_2 and the points x_1, x_2 are to be chosen so that the rule is exact when $f(x)$ is a cubic polynomial, $f(x) \equiv a_0 + a_1 x + a_2 x^2 + a_3 x^3$. Equating coefficients, we obtain four equations

$$[a_0] \quad w_1 + w_2 = \int_{-1}^{1} dx = 2$$

$$[a_1] \quad w_1 x_1 + w_2 x_2 = \int_{-1}^{1} x \, dx = 0$$

$$[a_2] \quad w_1 x_1^2 + w_2 x_2^2 = \int_{-1}^{1} x^2 \, dx = 2/3$$

$$[a_3] \quad w_1 x_1^3 + w_2 x_2^3 = \int_{-1}^{1} x^3 \, dx = 0$$

The equations for a_1 and a_3 are satisfied by $w_1 = w_2$ and $x_1 = -x_2$. From $[a_0]$, $w_1 + w_2 = 2 \Rightarrow w_1 = w_2 = 1$, and from $[a_2]$, $w_1 x_1^2 + w_2 x_2^2 = 2/3 \Rightarrow x_1^2 = x_2^2 = 1/3$. Hence the 2-point Gaussian rule is given by

$$\int_{-1}^{1} f(x) \, dx = f(1/\sqrt{3}) + f(-1/\sqrt{3}) + E_2^{(G)} \quad \square$$

In principle we could determine the weights and points for any positive integer p by solving the $2p$ simultaneous equations which arise from equating coefficients in a polynomial of degree $2p - 1$. However, the equations are non-linear, and the derivation becomes very complicated even for small values of p.

Fortunately, there is an elegant mathematical technique which yields the weights and points quite easily and also provides an expression for the truncation error. This technique is based on the theory of *orthogonal polynomials*.

Definition 8.1

A set of polynomials $P = \{p_n(x) \mid n = 0, 1, 2, \ldots\}$, where n denotes the degree, is said to be *orthogonal* with respect to a weight function $\omega(x) \geqslant 0$ over an interval $[a, b]$ if

$$\int_a^b \omega(x)\, p_m(x)\, p_n(x)\, dx = 0 \qquad \text{for } m \neq n \quad \square$$

There are several important sets of orthogonal polynomials, depending on the choice of the weight function $\omega(x)$ and the interval $[a, b]$. For $\omega(x) \equiv 1$ and $a = -1, b = 1$, the polynomials are known as *Legendre polynomials*. The first is $p_0^{(L)}(x) \equiv 1$, and the next three are

$$p_1^{(L)}(x) \equiv x \quad \text{(zero: 0)}$$

$$p_2^{(L)}(x) \equiv (3x^2 - 1)/2 \quad \text{(zeros: } \pm 1/\sqrt{3})$$

$$p_3^{(L)}(x) \equiv (5x^3 - 3x)/2 \quad \text{(zeros: } 0, \pm\sqrt{(3/5)})$$

The Legendre polynomials satisfy the recurrence relation

$$p_{n+1}^{(L)}(x) = \left(\frac{2n+1}{n+1}\right) x\, p_n^{(L)}(x) - \left(\frac{n}{n+1}\right) p_{n-1}^{(L)}(x), \quad n \geqslant 1$$

We state the following theorem without proof. A proof can be found in many textbooks of numerical analysis; a particularly clear account is given in Cohen (1973), chapter 5.

Theorem 8.2

The integration points x_i for the p-point Gaussian rule† (8.11) are the zeros of the Legendre polynomial of degree p, and the weights are given by

$$w_i = \frac{1}{[p_p^{(L)\prime}(x_i)]^2} \times \frac{2}{(1 - x_i^2)}, \quad i = 1, 2, \ldots, p$$

Further, the truncation error of the p-point Gaussian rule is

$$E_p^{(G)} = \frac{2^{2p+1}\,(p!)^4}{(2p+1)\,[(2p)!]^3}\, f^{(2p)}(\xi), \quad \xi \in (-1, 1) \quad \square$$

---

†Since the integration points for the rule in equation 8.11 are the zeros of a Legendre polynomial, the rule is often called a *Gauss–Legendre rule*.

From theorem 8.2 we can obtain the parameters for the 1-point Gaussian rule. The integration point is given by the zero of the polynomial $p_1^{(L)}(x) \equiv x$; thus $x_1 = 0$. The weight is

$$w_1 = \frac{1}{1^2} \times \frac{2}{1} = 2$$

and the truncation error is

$$E_1^{(G)} = \frac{2^3}{3} \times \frac{1^4}{2^3} f''(\xi) = \frac{1}{3} f''(\xi), \quad \xi \in (-1, 1)$$

Similarly, the truncation error of the 2-point Gaussian rule is

$$E_2^{(G)} = \frac{2^5}{5} \times \frac{2^4}{(24)^3} f^{(4)}(\xi) = \frac{1}{135} f^{(4)}(\xi), \quad \xi \in (-1, 1)$$

The accuracy of the 2-point Gaussian rule is directly comparable with that of Simpson's rule (see exercise 8.7); the advantage of the Gaussian rule is that it involves only two evaluations of the integrand.

Gaussian rules are examples of *open* integration formulae; in contrast with the *closed* Newton–Cotes formulae in equation 8.8, all the integration points lie within the range of integration and the end-points of the range are not included.

Procedure gaussint

Theorem 8.2 enables us to compute the weights and integration points for the p-point Gaussian rule. These parameters are tabulated in a number of texts, for example Abramowitz and Stegun (1965) where the weights and points for various values of p up to 96 are recorded. We provide the following procedure to set up the weights and points for the first four Gaussian rules, first defining appropriate data types.

```
TYPE
    posint = 1 .. maxint ;
    gaussrules = 1 .. 4 ;
    gvector = ARRAY [gaussrules] OF real ;


PROCEDURE gsetup ( p : gaussrules ;
                   VAR weight, point : gvector ) ;

    VAR
        i : gaussrules ;

    BEGIN
    CASE p OF
        1:  BEGIN
            weight[1] := 2.0 ;
            point [1] := 0.0
            END ;
        2:  BEGIN
            weight[2] := 1.0 ;
            point [2] := 0.5773502692
            END ;
```

```
    3:   BEGIN
         weight[2] := 0.8888888889 ;
         weight[3] := 0.5555555556 ;
         point [2] := 0.0 ;
         point [3] := 0.7745966692
         END ;
    4:   BEGIN
         weight[3] := 0.6521451549 ;
         weight[4] := 0.3478548451 ;
         point [3] := 0.3399810436 ;
         point [4] := 0.8611363116
         END
    END (* CASE *) ;

  FOR i := 1 TO p DIV 2 DO
     BEGIN
     weight[i] :=  weight[p+1-i] ;
     point [i] := -point [p+1-i]
     END
  END (* gsetup *)
```

We shall use procedure gsetup in a routine called gaussint to be developed
for the mathlib library. The library routine approximates the integral
$I = \int_a^b f(x) \, dx$ by a *composite* Gaussian rule. The user selects a particular Gaussian
rule by supplying a parameter p with an integer value from 1 to 4. The user must
also provide a positive integer n which is used to partition $[a, b]$ into n subintervals of equal width; thus $x_0 = a$, $x_i = x_0 + ih$ for $i = 0, 1, \ldots, n$ where
$h = (b - a)/n$.

We write the integral as

$$I = \int_a^b f(x) \, dx = \sum_{i=1}^n \int_{x_{i-1}}^{x_i} f(x) \, dx$$

and apply the p-point Gaussian rule to each subintegral. First, however, we must
transform the interval of integration from $[x_{i-1}, x_i]$ to $[-1, 1]$. For the integral

$$\int_{x_\varrho}^{x_r} f(x) \, dx \tag{8.12}$$

we make the linear transformation

$$x = \frac{(x_r - x_\varrho)}{2} y + \frac{(x_r + x_\varrho)}{2}$$

and obtain

$$\int_{-1}^1 f\left(\frac{(x_r - x_\varrho)}{2} y + \frac{(x_r + x_\varrho)}{2}\right) \frac{(x_r - x_\varrho)}{2} \, dy$$

We can now apply our Gaussian rule to this integral. The following Pascal
function computes integral 8.12 by this method; the weights and points for the
Gaussian rule, which are global to the function, must be set up previously by a
call of gsetup.

```
FUNCTION gsubint ( xl, xr : real ) : real ;
(* Computes the integral of f(x) over [xl, xr]
   using a p-point Gaussian rule *)

   VAR
      m, c, sum : real ;
      i           : gaussrules ;

   BEGIN
   (* A linear mapping x = m*y + c is defined
      from [-1, 1] to [xl, xr] *)
   m := (xr - xl)/2 ;  c := (xr + xl)/2 ;
   sum := 0.0 ;
   FOR i := 1 TO p DO
      sum := sum + weight[i] * f(m*point[i] + c) * m ;
   gsubint := sum  (* estimate of integral *)
   END (* gsubint *)
```

We can now write our library routine gaussint, incorporating gsetup and
gsubint.

```
PROCEDURE gaussint (
   FUNCTION f(x:real) : real ;        (* specification of integrand *)
               a, b   : real ;        (* limits of integration       *)
               p      : gaussrules ;  (* number of points in rule *)
               n      : posint ;      (* number of subintervals      *)
          VAR ivalue: real            (* computed value of integral *)
                         ) ;

(* This procedure estimates the integral of f(x) from x=a to x=b
   using a composite p-point Gaussian rule, where p is an integer
   from 1 to 4 selected by the user. The interval [a, b] is
   partitioned into n subintervals of equal width, and the
   integral of f over each subinterval is estimated using the
   p-point Gaussian rule.  The computed value of the integral is
   returned in ivalue. There are no error numbers and no
   indicator of accuracy in gaussint *)

   TYPE
      gvector = ARRAY [gaussrules] OF real ;

   VAR
      sum, xl, xr, h : real ;
      i              : posint ;
      weight, point  : gvector ;

   <PROCEDURE gsetup> ;

   <FUNCTION gsubint> ;

   BEGIN  (* body of gaussint *)
   gsetup (p, weight, point) ;
   h := (b-a)/n ;
   sum := 0.0 ;
   FOR i := 1 TO n DO
      BEGIN
      xr := a + i*h ;  xl := xr - h ;
      sum := sum + gsubint(xl, xr)
      END ;
   ivalue := sum  (* estimate of the integral over [a, b] *)
   END (* gaussint *)
```

Notes

(i)   No estimate of the truncation error is provided in gaussint. It is possible to estimate the error by comparing the values computed for the integral using n and 2n subintervals (see exercise 8.7).

(ii)   Since the end-points *a* and *b* are not used in Gaussian rules, the routine can be applied to cases where the integrand is singular at one or both end-points (provided, of course, that the integral is defined). For example, gaussint could be used to estimate the integral

$$\int_0^1 \ln x \cos x \, dx$$

(However, the rate of convergence would be slow because of the singularity in the integrand and its derivatives). A closed formula such as Simpson's rule could not be used in such cases.

(iii) It would be more efficient to partition the interval $[a, b]$ in such a way that a small stepwidth is used where the integrand is changing rapidly and a larger stepwidth elsewhere. It would be difficult and inconvenient for the user to supply such a partitioning, and in the library routine we have simply taken a fixed stepwidth. In the next chapter we shall see how a variable stepwidth can be determined automatically.

8.4   COMPARISON OF INTEGRATION RULES

We have seen that there are a number of different rules (indeed different families of rules) which can be used to approximate the value of a definite integral. We are now faced with the problem of distinguishing between these rules and selecting the best one to use in particular cases. It requires a mixture of experience, analysis (investigation of truncation error) and numerical testing to judge the best rule to use for any given integral.

In this context an important question is how efficient is the rule, that is, how much computation is required for a given accuracy? To answer this question we record the number of evaluations of the integrand and plot this number against the accuracy attained (evaluation of the integrand is usually by far the most important component in the computational cost). The resulting graph is called a *cost curve*. We plot the cost curves for several rules and compare them to decide which rule is most efficient for a particular type of integral.

Example 8.4

Draw the cost curves for the integral

$$I = \int_1^2 (e^{-x}/x) \, dx$$

using the composite trapezoidal rule, composite Simpson's rule and composite p-point Gaussian rules with $p = 1, 2, 3, 4$. Which rule is most efficient if an accuracy of 9 decimal places is required?

The exact value of the integral is known (see example 8.1). The accuracy obtained with the trapezoidal and Simpson's rule is tabulated in example 8.1 and section 8.2; the number of evaluations of the integrand is equal to $n + 1$, where n is the number of subintervals. Similar results for the Gaussian rules can be obtained from gaussint. The cost curve for each rule is drawn in figure 8.3. We see that the most efficient rule for an accuracy of 9 decimal places is the 4-point Gaussian rule.

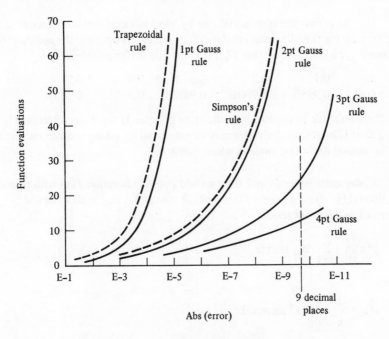

Figure 8.3   Cost curves for $I = \int_1^2 (e^{-x}/x)\,dx$

To carry out a serious comparison of the integration rules, we would of course need to experiment on a variety of different integrals. Nevertheless, the cost curves in figure 8.3 indicate that (at least if the integrand is smooth) the higher-order rules are most efficient when high accuracy is required. Also, the Gaussian rules are more efficient than the corresponding Newton–Cotes rules over a wide range of accuracies. These remarks are generally true if the integrand is well-behaved and sufficiently differentiable in the interval $[a, b]$. For low accuracies and cases where the integrand is not highly differentiable, the picture is less clear; we must then experiment to determine which rule is most appropriate in any particular case.

EXERCISES

8.1   Write a program using procedure trapint to estimate the integral

$$\int_0^1 \frac{4}{1+x^2}\,\mathrm{d}x \quad (=\pi)$$

Obtain an error bound from the formula 8.5, and find a stepwidth sufficiently small for an accuracy of 6 decimal places. Hence compute π to this accuracy. Execute the program for $n = 1, 2, 4, \ldots, 128$, and comment on the behaviour of the error as a function of n.

8.2   Use the composite trapezoidal rule by hand or calculator to estimate $\int_{0.1}^{0.5} f(x)\,\mathrm{d}x$ for the following tabulated function. Also estimate the accuracy of the result (you may assume that $f(x)$ is arbitrarily differentiable).

x	0.1	0.2	0.3	0.4	0.5
$f(x)$	0.9950	0.9801	0.9553	0.9211	0.8776

Use Simpson's rule to approximate the same integral. Is there any advantage in doing this? (The fourth derivative may be estimated by taking the approximation for the second derivative twice in succession.)

8.3   Under quite general conditions an odd periodic function $f(x)$ with period 2π (that is, $f(-x) = -f(x) = -f(x + 2\pi k), k = \pm 1, \pm 2, \ldots$, for all x) can be expressed as a *Fourier series*

$$f(x) = \sum_{m=1}^{\infty} d_m \sin mx$$

where

$$d_m = \frac{2}{\pi} \int_0^{\pi} f(x) \sin mx \, \mathrm{d}x$$

Write a program to compute the Fourier coefficients d_m for such a function, for $m = 1, 2, \ldots, M$, correct to 4 decimal places. (Use procedure trapint; start with $n = 1$ and double n successively until the estimate of truncation error given by the formula in equation 8.6 is sufficiently small.)

Test the program on the periodic sawtooth function defined by $f(x) = x$ for $-\pi < x < \pi, f(\pi) = 0$, and compute the coefficients $d_1, \ldots, d_6$. Use the library routine graf to plot the Fourier approximation to the function over $[-2\pi, 2\pi]$ based on the coefficients you have calculated.

8.4   Verify that the weights for the Newton–Cotes formula in the case $p = 3$ give the *3/8 rule* as stated in section 8.2. Write a program to estimate the integral

$$\int_1^2 (e^{-x}/x)\,\mathrm{d}x$$

by a composite form of the 3/8 rule. Tabulate the absolute error and an error bound, draw a cost curve, and compare with the curves in figure 8.3.

8.5 The resistivity ρ of a semiconductor is measured over the temperature range $10, 20, \ldots, 100$ degrees C, and the following ten values are found (in ohm-metres): $0.124, 0.119, 0.115, 0.110, 0.107, 0.103, 0.100, 0.097, 0.094, 0.092$. The values are believed to be correct to 3 decimal places. Write a program to compute the integral of ρ with respect to temperature over this range, using a suitable library routine.

It is known on theoretical grounds that ρ is arbitrarily differentiable with respect to temperature and that the fourth derivative does not exceed 10^{-5} in absolute value over this range. Estimate the maximum possible error in the computed result arising from (i) truncation error, (ii) data error. How many figures in the result would you expect to be correct?

8.6 Write a program to compute double integrals of the form

$$\int_{x=a}^{b} \int_{y=c(x)}^{d(x)} f(x, y)\, dx\, dy$$

using procedure **gaussint**. (*Hint:* Express the integral as

$$I = \int_{a}^{b} g(x)\, dx$$

where

$$g(x) = \int_{y=c(x)}^{d(x)} f(x, y)\, dy$$

Write a Pascal function which invokes gaussint to estimate $g(x)$ for a specified (fixed) value of x. You may then call gaussint with the function in the main body of the program to estimate the outer integral I.)

Try your program on

(a) $\displaystyle\int_{0}^{1}\int_{0}^{1} \sin(x^2 y^2)\, dx\, dy$

(b) $\displaystyle\int_{1}^{2}\int_{0}^{x} e^{y/x}\, dx\, dy$

(c) $\displaystyle\int_{R}\int \sqrt{(1 + e^{xy})}\, dx\, dy$ where R is the region bounded by the unit circle.

8.7 Show by theorem 8.2 that the truncation error of the p-point Gaussian rule applied to the integral $\int_{0}^{h} f(x)\, dx$ is given by

$$c_p\, h^{2p+1}\, f^{(2p)}(\xi), \quad \xi \in (0, h)$$

provided f is sufficiently differentiable, where c_p is a constant which depends on p. Hence obtain an expression for the error of the composite p-point Gaussian rule used in procedure **gaussint**. If $G_p(n)$ and $G_p(2n)$ denote estimates of the integral $I = \int_a^b f(x)\,dx$ obtained by the composite Gaussian rule with n and $2n$ subintervals respectively, and f is sufficiently smooth, show that

$$I - G_p(2n) \approx \frac{G_p(2n) - G_p(n)}{2^{2p} - 1}$$

Repeat exercise 8.3 using **gaussint**, and determine experimentally the best value of p to use.

9 Numerical Integration: Adaptive Methods

They know not well the subtle ways
I keep, and pass, and turn again.   Ralph Waldo Emerson, *Brahma*

In chapter 8 we discussed some basic numerical integration rules and developed two simple routines for the mathlib library. These routines have the major disadvantage that no indicator of numerical error is provided. Also, in gaussint, the user is required to select a particular gaussian rule, which presupposes some understanding of how the rule works. Ideally we would like to have a library routine which requires only the information necessary to specify the problem (the integrand, the limits of integration and the error tolerance). Given this information, the routine should compute the integral to a specified accuracy, determining the stepsize and other parameters automatically, or warn the user if the accuracy cannot be achieved. Routines of this type are called *automatic integrators*.

Unfortunately, it is not possible to develop an automatic integrator based on just one integration rule which will be equally effective for the many different types of integral which arise in practice. For this reason commercial libraries provide a large selection of different integration routines†. In this chapter we shall consider in detail two approaches to the problem and develop two automatic integrators, called romberg and adsimp. These routines can be accessed individually in the mathlib library; they can also be called via a master integration routine integral which will be developed later in the chapter. Procedure integral incorporates a method of transforming difficult integrals into a form which can be handled effectively by one or both of our standard integrators; it can also be used to compute integrals over semi-infinite or infinite intervals.

9.1   OUTLINE OF AN AUTOMATIC INTEGRATOR BASED ON THE TRAPEZOIDAL RULE

To illustrate the main points in the design of an automatic integrator, we consider first a method based on the composite trapezoidal rule. We recall from section 8.1

†The NAG library (Mark 8) has nineteen routines for numerical integration

213

that the integral $I = \int_a^b f(x)\,\mathrm{d}x$ can be approximated by the rule

$$T(h) = h\left(\tfrac{1}{2}f_0 + f_1 + \ldots + f_{n-1} + \tfrac{1}{2}f_n\right)$$

where $h = (b - a)/n$. We also recall our remark that the truncation error for step-size $h/2$ can be estimated from the formula

$$I - T(h/2) \approx \frac{T(h/2) - T(h)}{3} \tag{9.1}$$

To derive the formula 9.1 we make use of the following theorem, which will be stated without proof (see, for example, Dahlquist and Björck (1974), chapter 7).

Theorem 9.1

If the integrand $f(x)$ has continuous derivatives up to $f^{(2k+2)}$ on $[a, b]$, then the truncation error of the composite trapezoidal rule can be expressed as

$$I - T(h) = b_2 h^2 + b_4 h^4 + b_6 h^6 + \ldots + b_{2k} h^{2k} + O(h^{2k+2})$$

where

$$b_{2i} = B_{2i}\left(f^{(2i-1)}(a) - f^{(2i-1)}(b)\right)$$

and the quantities B_{2i} are constants independent of h and f. □

The error expansion in theorem 9.1, which is known as the *Euler–Maclaurin formula*, can be used to derive the error estimate in formula 9.1. Assuming that f is at least four times continuously differentiable on $[a, b]$, we have

$$I - T(h) = b_2 h^2 + O(h^4)$$

The first term on the right is known as the *principal truncation error* because, when h is sufficiently small, this term dominates the expression for the truncation error. To estimate the principal truncation error we rewrite the equation for a stepsize $h/2$

$$I - T(h/2) = b_2 (h/2)^2 + O(h^4)$$

(here we have used the fact that the coefficient b_2 is independent of h). Eliminating the term in h^2 between the two equations, we obtain

$$4(I - T(h/2)) = I - T(h) + O(h^4)$$

hence

$$I - T(h/2) = \frac{T(h/2) - T(h)}{3} + O(h^4) \tag{9.2}$$

Looking at this from another point of view, we can write equation 9.2 as

$$I = \frac{4T(h/2) - T(h)}{3} + O(h^4) \tag{9.3}$$

The approximation $I \approx (4T(h/2) - T(h))/3$ is thus a fourth-order approximation to the value of the integral. In fact, it is easy to show that equation 9.3 gives the composite form of Simpson's rule for a stepwidth $h/2$, that is, $S(h/2) = (4T(h/2) - T(h))/3$. This method of eliminating the lowest power of h in the error expansion, using two different stepwidths such as h and $h/2$, is known as *Richardson's extrapolation*. The method is attractive because, once we have computed $T(h)$ and $T(h/2)$, we can obtain the higher-order approximation in equation 9.3 at virtually no extra cost.

Example 9.1

Estimate the integral $\int_0^1 e^{\sin x}\, dx$ using the composite trapezoidal rule with $h = 1, 1/2$ and $1/4$. Improve the estimates by Richardson's extrapolation.
   We obtain

h	$T(h)$	extrapolated values
1	1.65988841	
$\frac{1}{2}$	1.63751735	1.63006033
$\frac{1}{4}$	1.63321154	1.63177627

The extrapolated values are obtained from the formula in equation 9.3; for example, $(4T(\frac{1}{2}) - T(1))/3 = 1.63006033$. The exact value of the integral is $1.63186961\ldots$, and we see that the extrapolated values are a considerable improvement on those obtained from the trapezoidal rule. □

Design of an Automatic Integrator

An automatic integrator has three main components:

(i)   an integration rule (or sometimes two rules);
(ii)   a formula for estimating the truncation error;
(iii) a strategy for deciding when to terminate the calculation.

For (i) we shall use in this section the composite trapezoidal rule. We know that if the integrand is at least four times continuously differentiable on $[a, b]$, then when h is sufficiently small the error of the composite trapezoidal rule for stepsize $h/2$ is

$$I - T(h/2) \approx (T(h/2) - T(h))/3 \qquad (9.4)$$

This gives us a formula for estimating the error. Finally, we need a strategy for selecting an appropriate stepwidth. The simplest approach is to start with the maximum stepwidth, $h_0 = b - a$, and compute estimates $T(h_0)$ and $T(h_0/2)$. If formula 9.4 tells us that the error is too large, we continue halving the stepwidth until the error is acceptably small.
   We recall from section 8.1 that the trapezoidal estimates for stepwidths h and $h/2$ are related by the equation

$$T(h/2) = \tfrac{1}{2} T(h) + \frac{h}{2} \sum_{i=1}^{n} f(a + (i - \tfrac{1}{2})h)$$

Since we are halving the stepwidth in the automatic integrator, we can use this relation to economise on the number of function evaluations. This is implemented in the following procedure.

```
PROCEDURE traprule (     firstcall : boolean ;
                     VAR ivalue, h : real ;
                     VAR n         : posint  ) ;
(* Estimates the integral of f(x) from x=a to x=b using
   the composite trapezoidal rule. The function f and the
   limits a and b are global to the procedure. If on entry
   firstcall=true, then the trapezoidal rule is applied with
   h = b-a, n=1. Otherwise it is assumed that on entry ivalue
   contains the trapezoidal estimate of the integral for a
   given stepwidth h. On exit the stepwidth is halved, and
   ivalue contains the new trapezoidal estimate for stepwidth
   h/2. The parameter n records the current number of
   subintervals of the interval [a, b] *)

VAR
   newsum : real ;
   i      : posint ;

BEGIN
IF firstcall THEN
   BEGIN
   h := b-a ;  n := 1 ;
   ivalue := (h/2) * (f(a) + f(b))
   END
ELSE
   BEGIN
   newsum := 0.0 ;
   FOR i := 1 TO n DO
      newsum := newsum + f(a + (i-0.5)*h) ;
   ivalue := ivalue/2 + (h/2)*newsum ;
   h := h/2 ;  n := 2*n
   END
END (* traprule *)
```

We now outline an algorithm which, under suitable conditions on the integrand f, will automatically estimate the value of an integral $\int_a^b f(x)\, \mathrm{d}x$ to a specified absolute error tolerance tol.

```
BEGIN  (* automatic integrator based on the trapezoidal rule *)
Specify f, a, b and tol ;
(* Start integration with whole interval [a, b] *)
traprule (true, ivalue, h, n) ;
REPEAT
   lastivalue := ivalue ;
   traprule (false, ivalue, h, n) ;
   errest := (ivalue - lastivalue) / 3
UNTIL abs(errest) < tol ;
(* Apply Richardson's extrapolation *)
ivalue := ivalue + errest  (* final estimate of the integral *)
END .
```

Notes

(i)   The number of evaluations of f depends on the number of subintervals of $[a, b]$, which in turn depends on the value specified for tol. If the algorithm terminates after k iterations with a stepsize $h_k = (b - a)/2^k$, a total of $2^k + 1$ function evaluations are performed.

(ii)   Once the required accuracy has been attained, we apply Richardson's extrapolation using the formula in equation 9.2. The final estimate therefore has fourth-order accuracy provided that f is at least four times continuously differentiable on $[a, b]$. However, because we are controlling the truncation error in the trapezoidal rule, which is $0(h^2)$, the algorithm is said to be a second-order process.

(iii)   To simplify the coding of our integration routines we shall control the truncation error in all cases by an absolute error test. In many libraries a more refined test is provided; a typical form would be

$\quad$ |I − ivalue| < tol

where

$\quad$ tol = max (epsabs, epsrel · | ivalue|)

The parameters epsabs and epsrel are set by the user. When epsrel = 0 we have an absolute error test of the form used in mathlib; when epsabs = 0 we have a relative error test, which can be used to measure the number of correct significant digits in ivalue (particularly useful when I is large or small in absolute value). If the magnitude of I is not known in advance, it is common practice to use a *mixed error test*, setting epsabs = epsrel.

Before we write a library routine for automatic integration, we must take account of the requirements of adaptability and robustness; for example, an impossibly small tolerance may be set which would require a stepwidth too small for the computer. Another problem is that our error estimates are necessarily based on the assumption that the integrand is sufficiently differentiable. For many integrals this is not the case; a simple example is $\int_{-1}^{1} |x|\, dx$. In general, the sequence $\{T(h_k)\}, k = 0, 1, \ldots,$ where $h_k = (b - a)/2^k$, will converge to I (provided I is defined), but the rate of convergence may be very slow.

It is possible to devise integrals which deceive automatic integrators. This normally means that the integration process terminates before the required accuracy has been attained. It is easy to fool our integrator based on the composite trapezoidal rule. For the integral

$$I = \int_0^1 e^{-x} \sin 2\pi x \, dx = \frac{2\pi}{1 + (2\pi)^2} (1 - e^{-1})$$

we find that

$$T(1) = \tfrac{1}{2} [f(0) + f(1)] = \tfrac{1}{2}(\sin 0 + e^{-1} \sin 2\pi) = 0$$

and

$$T(\tfrac{1}{2}) = \tfrac{1}{2}T(1) + \tfrac{1}{2}f(\tfrac{1}{2}) = \tfrac{1}{2}(0 + e^{-\frac{1}{2}} \sin \pi) = 0$$

Hence errest = 0, and the process terminates with the incorrect result ivalue = 0. To minimise the chance of this happening in general, we often perform two or three iterations before checking to see if the error test is satisfied.

9.2 ROMBERG INTEGRATION

We could develop a reliable and robust automatic integrator based on the algorithm outlined in section 9.1. However, as we see from the cost curves in figure 8.3, the composite trapezoidal rule converges relatively slowly, and this means that the method is very inefficient when high accuracy is required. We would like to develop an algorithm which selects not only the stepwidth but also the order of the integration rule. Such a method is available: it is called *Romberg integration*.

The underlying idea of Romberg integration is to apply Richardson's extrapolation *repeatedly* to eliminate successive powers of h in the error expansion of theorem 9.1,

$$I - T(h) = b_2 h^2 + b_4 h^4 + b_6 h^6 + \ldots \tag{9.5}$$

(where the coefficients b_{2i} are independent of h). Eliminating the term in h^2, as in equations 9.2 and 9.3, we obtain the extrapolated value

$$R^{(1)}(h) = \frac{4T(h/2) - T(h)}{3}$$

which satisfies

$$I - R^{(1)}(h) = \overline{b}_4 h^4 + \overline{b}_6 h^6 + \ldots \tag{9.6}$$

(the coefficients $\overline{b}_4, \overline{b}_6, \ldots$ are again independent of h). This means that there exists a positive number C such that, for h sufficiently small

$$|I - R^{(1)}(h)| \leqslant Ch^4$$

From equation 9.5 and the corresponding equation for $I - T(h/2)$, we can derive

$$T(h/2) - T(h) = \tfrac{3}{4} b_2 h^2 + 0(h^4)$$

As $h \to 0$ the term in h^2 dominates the right hand side; therefore, provided that h is sufficiently small, we have

$$|T(h/2) - T(h)|/3 \geqslant Ch^4$$

It follows that, for h sufficiently small,

$$|I - R^{(1)}(h)| \leqslant |T(h/2) - T(h)|/3$$
$$= |R^{(1)}(h) - T(h/2)| \tag{9.7}$$

The formula 9.7 provides a means of estimating an error bound for the extrapolated value $R^{(1)}(h)$. If the required accuracy is not attained at this stage, we can halve the stepwidth, compute $T(h/4)$, and extrapolate to find

$$R^{(1)}(h/2) = \frac{4T(h/4) - T(h/2)}{3}$$

We may then use $R^{(1)}(h)$ and $R^{(1)}(h/2)$ to eliminate the term in h^4 from equation 9.6. We have

$$I - R^{(1)}(h) = \bar{b}_4 h^4 + 0(h^6)$$
$$I - R^{(1)}(h/2) = \bar{b}_4 (h/2)^4 + 0(h^6)$$

hence

$$2^4 [I - R^{(1)}(h/2)] = I - R^{(1)}(h) + 0(h^6)$$

and finally

$$I = \frac{2^4 R^{(1)}(h/2) - R^{(1)}(h)}{2^4 - 1} + 0(h^6)$$

We define

$$R^{(2)}(h) = \frac{2^4 R^{(1)}(h/2) - R^{(1)}(h)}{2^4 - 1}$$

From equation 9.6 we can obtain

$$R^{(1)}(h/2) - R^{(1)}(h) = \tfrac{15}{16} \bar{b}_4 h^4 + 0(h^6)$$

A similar argument to the one used above then shows that, for h sufficiently small,

$$|I - R^{(2)}(h)| \leqslant |R^{(1)}(h/2) - R^{(1)}(h)| / 15$$

$$= |R^{(2)}(h) - R^{(1)}(h/2)| \tag{9.8}$$

This is an approximate error bound for $R^{(2)}(h)$ corresponding to equation 9.7 for $R^{(1)}(h)$.

The Romberg Table

It is convenient to set out the extrapolated values in tabular form as follows:

$T(h)$
$T(h/2)$ $\quad R^{(1)}(h)$
$T(h/4)$ $\quad R^{(1)}(h/2)$ $\quad R^{(2)}(h)$

This is called a *Romberg table*. Provided the integrand is at least six times continuously differentiable on $[a, b]$, the error of the final estimate $R^{(2)}(h)$ is $0(h^6)$.

The process can be continued. The ith row of the Romberg table is of the form

$$T\left(\frac{h}{2^{i-1}}\right) \quad R^{(1)}\left(\frac{h}{2^{i-2}}\right) \quad \cdots \quad R^{(i-2)}\left(\frac{h}{2}\right) \quad R^{(i-1)}(h)$$

Provided the integrand is sufficiently differentiable, we have

$$I - R^{(j)}(h) = \hat{b}_{2j+2}\, h^{2j+2} + 0(h^{2j+4}), \quad j = 1, 2, \ldots$$

where the coefficients $\hat{b}_{2j+2}$ are independent of h. Hence for a halved stepwidth we can write

$$I - R^{(j)}(h/2) = \hat{b}_{2j+2}\left(\frac{h}{2}\right)^{2j+2} + 0(h^{2j+4})$$

We may now, following the same procedure as before, eliminate the term in h^{2j+2} to obtain

$$I = R^{(j+1)}(h) + 0(h^{2j+4})$$

where

$$R^{(j+1)}(h) = \frac{2^{2j+2}\, R^{(j)}(h/2) - R^{(j)}(h)}{2^{2j+2} - 1}$$

$$= R^{(j)}(h/2) + \frac{R^{(j)}(h/2) - R^{(j)}(h)}{2^{2j+2} - 1} \tag{9.9}$$

We can also show that, for h sufficiently small

$$|I - R^{(j+1)}(h)| \leqslant \frac{|R^{(j)}(h/2) - R^{(j)}(h)|}{2^{2j+2} - 1}$$

$$= |R^{(j+1)}(h) - R^{(j)}(h/2)| \tag{9.10}$$

With the convention that $R^{(0)}(h) = T(h)$, the formulae in equations 9.9 and 9.10 apply for $j = 0, 1, 2, \ldots, i - 2$ on the ith row of the Romberg table.

Example 9.2

Estimate the integral $\int_0^1 e^{\sin x}\, dx$ by Romberg integration.

h	$T(h)$	$R^{(1)}$	$R^{(2)}$	$R^{(3)}$
1	1.65988841			
$\frac{1}{2}$	1.63751735	1.63006033		
$\frac{1}{4}$	1.63321154	1.63177627	1.63189067	
$\frac{1}{8}$	1.63220091	1.63186403	1.63186988	<u>1.63186955</u>

The integrand is arbitrarily differentiable on $[0, 1]$. The extrapolated values are calculated from equation 9.9 with $j = 0, 1, 2$; the final estimate, which is an $0(h^8)$ approximation, is 1.63186955. The estimated error bound, from equation 9.10, is

$$|R^{(3)}(h) - R^{(2)}(h/2)| = |1.63186955 - 1.63186988|$$
$$= 3.3 \times 10^{-7}$$

(the actual error is 0.6×10^{-7}). □

Romberg integration in a table of r rows is applicable if the integrand f has continuous derivatives up to $f^{(2r)}$ on $[a, b]$. As we have seen, the values in column j then have $0(h^{2j+2})$ convergence for $0 \leqslant j \leqslant r - 1$. When Romberg integration is applicable, it is highly efficient. In example 9.2 only nine evaluations of the integrand are required, whereas over twenty evaluations would be required in Simpson's rule and several hundred in the composite trapezoidal rule for the same accuracy.

9.3   IMPLEMENTATION OF ROMBERG INTEGRATION

We can now give an outline description of an automatic integrator based on Romberg integration. To simplify the description we shall use the following notation for the Romberg table

$T(h)$	$R^{(1)}$	$R^{(2)}$ . . .
$T_{1,0}$		
$T_{2,0}$	$T_{2,1}$	
$T_{3,0}$	$T_{3,1}$	$T_{3,2}$
.	.	. .
.	.	. .
.	.	. .

From equation 9.9 we have

$$T_{i,j+1} = T_{i,j} + (T_{i,j} - T_{i-1,j})/(4^{j+1} - 1) \tag{9.11}$$

for $i = 2, 3, \ldots, r$ and $j = 0, 1, \ldots, i - 2$. We shall use this to construct a new row of the Romberg table at each iteration. From 9.10 we have an error bound

$$|I - T_{i,j+1}| \leqslant |T_{i,j+1} - T_{i,j}| \tag{9.12}$$

provided h is sufficiently small. We shall use this to test for convergence when a new row of the table has been constructed.

The current row of the Romberg table is stored in a *vector* rtable. For the ith row

$$\text{rtable}[j] = T_{i,j}, \, j = 0, 1, \ldots, p$$

where

$$p = \min(\text{maxcol}, i - 1)$$

The number of columns in the table is restricted by a constant maxcol, which has the effect of limiting the maximum order of the integration process; a suitable value for maxcol is normally between 5 and 10. The following algorithm is based on procedure traprule from section 9.1, which is used to compute rtable[0]. When a complete new row has been constructed, the latest estimate of the integral is contained in rtable[p] and the estimate of the error bound is given by abs(rtable[p] − rtable[p − 1]).

```
BEGIN  (* automatic integrator based on Romberg integration *)
Specify f, a, b and tol ;
(* Call traprule to compute and store initial estimate *)
traprule (true, rtable[0], h, n) ;
p := 0 ;
REPEAT
   IF p < maxcol THEN
      p := p + 1 ;
   (* Call traprule to halve the stepwidth and start next row of
      the Romberg table with an improved trapezoidal estimate *)
   rj := rtable[0] ;
   traprule (false, rj, h, n) ;
   powerof4 := 4 ;
   (* Construct a new row of the Romberg table *)
   FOR j := 0 TO p-1 DO
      BEGIN
      diff := rj - rtable[j] ;
      rtable[j] := rj ;
      (* Compute next extrapolated value *)
      rj := rj + diff / (powerof4 - 1) ;
      powerof4 := 4*powerof4
      END ;
   rtable[p] := rj ;
   errbnd := abs (rj - rtable[p-1])
UNTIL errbnd < tol ;
ivalue := rtable[p]  (* final estimate of the integral *)
END .
```

Cautious Extrapolation

As we have seen in example 9.2, Romberg integration can be very efficient. However, great care is needed, because the extrapolation process is valid only when the integrand f is sufficiently differentiable. This is quite a severe restriction. For example, the apparently harmless integral $\int_0^1 \sqrt{(1-x^2)}\, e^x \, dx$ leads at once to

$$f'(x) = \sqrt{(1-x^2)}\, e^x - \frac{x}{\sqrt{(1-x^2)}}\, e^x$$

which has a singularity at $x = 1$. The conditions of theorem 9.1 are not satisfied, so we cannot assume that the Euler–Maclaurin expansion is valid even to the term in h^2. Romberg integration is therefore unlikely to work satisfactorily for this integral.

We must be more cautious about how we apply Richardson's extrapolation.

If we assume that f has continuous derivatives up to $f^{(2j+4)}$ on $[a, b]$, then from the analysis in section 9.2 we have

$$I - R^{(j)}(h) = \hat{b}_{2j+2} h^{2j+2} + 0(h^{2j+4})$$

where $\hat{b}_{2j+2}$ is a constant independent of h. We can be sure that the next extrapolation stage will produce an improved estimate only if the first term on the right dominates the error. That is, we require

$$I - R^{(j)}(h) \approx \hat{b}_{2j+2} h^{2j+2}$$

and so, by the usual step-halving process

$$R^{(j)}(h/2) - R^{(j)}(h) \approx \hat{b}_{2j+2} h^{2j+2} (1 - 1/2^{2j+2})$$

This relation leads to the *convergence ratio*

$$C_R^{(j)} = \frac{R^{(j)}(h) - R^{(j)}(2h)}{R^{(j)}(h/2) - R^{(j)}(h)} \approx 2^{2j+2} = 4^{j+1} \tag{9.13}$$

We are justified in accepting the extrapolated value given by equation 9.9 and relying on the error bound in equation 9.10 only if $C_R^{(j)} \approx 4^{j+1}$. This condition is the essence of *cautious extrapolation*, which is used in the integrator CADRE written by C. de Boor (described in Rice (1971)). The resulting process is less efficient than normal Romberg integration because three values are required before we can extrapolate, but we can be more confident that our integrator will return reliable results.

The first stage of cautious extrapolation is to test that the trapezoidal estimates have $0(h^2)$ convergence. We compute the ratio

$$C_R^{(0)} = \frac{T(h/2) - T(h)}{T(h/4) - T(h/2)}$$

for $h = b - a$, $(b - a)/2$, ..., and apply the test

$$|C_R^{(0)} - 4| < \text{h2tol}$$

A suitable value for the constant h2tol can only be found by numerical experiment; following de Boor we shall take h2tol = 0.15. If the test is satisfied, we extrapolate and test the convergence ratio for the next column in the Romberg table. In general, we consider a segment of column j of the table (we use the notation introduced at the beginning of this section):

$$\vdots$$

$$T_{i-2,j}$$
$$T_{i-1,j}$$
$$T_{i,j}$$

$$\vdots$$

We compute the ratio

$$C_R{}^{(j)} = \frac{T_{i-1,j} - T_{i-2,j}}{T_{i,j} - T_{i-1,j}}$$

for $i \geqslant j + 3$ and apply a test based on equation 9.13,

$$|C_R{}^{(j)} - 4^{j+1}| < 4^{j+1} \text{ extol} \tag{9.14}$$

Again a suitable value for the constant extol is found by experiment. If the test is satisfied, we accept the extrapolated value $T_{i,j+1}$ given by equation 9.11 and the error bound 9.12.

It is possible to relax the convergence condition slightly. The values in column j may be converging but not fast enough to satisfy inequality 9.14. We detect and handle this by the following procedure, which applies to row i of the Romberg table.

$$E^{(1)} := 4;$$

For $j := 1, 2, \ldots, p - 2$:

Accept the extrapolated value $T_{i,j+1}$

if $C_R{}^{(j)} > E^{(j)}$, otherwise stop;

$E^{(j+1)} := 4E^{(j)}$ if $|C_R{}^{(j)} - 4E^{(j)}| < 4E^{(j)}$. extol

$:= E^{(j)}$ otherwise;

Take as an error bound for $T_{i,j+1}$: $|T_{i,j} - T_{i-1,j}|/(E^{(j+1)} - 1)$.
With $j = 1$, for example, if we find that $C_R{}^{(1)} > 4$ but the condition $|C_R{}^{(1)} - 4^2|$ $< 4^2 \times$ extol is not satisfied, we accept the extrapolated value $T_{i,2} = T_{i,1} +$ $(T_{i,1} - T_{i-1,1})/(4^2 - 1)$ given by equation 9.11 but take a more pessimistic view of the error. Instead of the usual error bound $|T_{i,1} - T_{i-1,1}|/(4^2 - 1)$, we take $|T_{i,1} - T_{i-1,1}|/(4 - 1)$. For the purposes of the relaxed convergence test we find that a suitable value for the constant extol is 0.1.

Finally, if we cannot detect $O(h^2)$ convergence in the trapezoidal estimates, this does not necessarily mean that the estimates are not converging. Provided that the order of convergence is higher than $O(h)$, we can use the difference $|T_{i,0} - T_{i-1,0}|$ as a bound on the truncation error of $T_{i,0}$. We apply the special test

$$\frac{|C_R{}^{(0)} - \bar{C}_R{}^{(0)}|}{|C_R{}^{(0)}|} < 0.1 \quad \text{(provided that } C_R{}^{(0)} \neq 0) \tag{9.15}$$

where $\bar{C}_R{}^{(0)}$ denotes the value of $C_R{}^{(0)}$ on row $i - 1$. If this condition is satisfied, and also $C_R{}^{(0)} > 2.0$, we decide that the order of convergence is higher than $O(h)$ and take the error bound above. The constants used here are obtained from numerical tests on the integrator CADRE; further details and justification can be found in the article by C. de Boor in Rice (1971).

Procedure romberg

We shall now describe the library procedure romberg which implements Romberg integration with cautious extrapolation. Since this routine is longer than any previously written for the mathlib library, we first describe the subprograms used in romberg before giving the main body of the procedure. The heading and declarations are

```
PROCEDURE romberg
  ( FUNCTION f(x:real) : real ;     (* specification of integrand *)
             a, b       : real ;    (* interval of integration    *)
             tol        : real ;    (* absolute error tolerance    *)
             trace      : boolean ;(* when trace=true the Romberg
                                     table is written to output *)
        VAR ivalue       : real ;    (* computed value of integral *)
        VAR errbnd       : real ;    (* estimated error bound      *)
        VAR eflag        : integer ) ;

(* This procedure estimates the integral of f(x) from x=a to x=b
   using Romberg integration with cautious extrapolation. The
   computed value of the integral is returned in ivalue. The
   process terminates normally when the estimated error bound,
   errbnd, is less than tol. If a reliable error bound cannot be
   determined, errbnd is returned as 1/smallreal and eflag is set
   to 2. If trace=true the Romberg table and convergence ratios
   are written to the output file.
   The error numbers for eflag are:
     =1  The required accuracy has not been attained -
         either limiting precision of the computer has been
         reached or the maximum 15 rows of the Romberg table
         have been used. An estimated error bound is still
         provided but this lies outside the specified
         tolerance  (warning)
     =2  No reliable error bound has been determined  (fatal).
   Routines called from the library:
   errormessage, tolmin, stepmin      *)

CONST
     maxrow = 15 ; (* maximum number of rows in Romberg table *)
     maxcol = 10 ; (* maximum number of extrapolation columns
                      in Romberg table                        *)
     limacc = 100; (* factor used in test to terminate the
                      iteration before rounding errors can
                      seriously affect the Romberg process    *)

TYPE
     widthoftable = 0 .. maxcol ;
     rowvector = ARRAY [widthoftable] OF real ;

VAR
     h, smallh, tola,
           extrpconst, lastcr0   : real ;
     j, p                        : widthoftable ;
     row                         : 1 .. maxrow ;
     n                           : posint ;
     rtable, diff, cratio        : rowvector ;
     limith, converged, convtest : boolean ;
```

Notes

(i)   The maximum number of rows and extrapolation columns in the Romberg table is restricted by maxrow and maxcol respectively. The actual values are fairly arbitrary; it is unlikely that maxcol will be reached, and maxrow is normally only reached if the trapezoidal estimates appear to have lower than $0\,(h^2)$ convergence.

(ii)   There are three vectors of type rowvector: rtable contains the current row $T_{i,j}, j = 0, 1, \ldots, p$;   diff contains the differences $T_{i,j} - T_{i-1,j}, j = 0, 1, \ldots, p - 1$;   cratio contains the convergence ratios $C_R^{(j)}, j = 0, 1, \ldots, p - 2$.

(iii)   A boolean parameter trace is provided so that the user can inspect the Romberg table and convergence ratios. This is often helpful in assessing the suitability of Romberg integration.

The basic step of the algorithm is to compute a new row of the Romberg table and to update the convergence ratios. This is done in the following procedure, which should be compared with the outline algorithm given at the beginning of this section.

```
PROCEDURE rombstep ( p : widthoftable ;
                VAR rtable, diff, cratio : rowvector ;
                VAR h : real ;  VAR n : posint ) ;
(* Constructs a new row of the Romberg table and computes
   the corresponding differences and convergence ratios.
   Invokes traprule to start the new row with an improved
   trapezoidal estimate. The vectors rtable, diff and cratio
   contain respectively the current row, the differences and
   the convergence ratios. If the global variable trace=true,
   the contents of rtable and cratio are written to the
   output file via the procedure traceoftable *)

VAR
   rj, newdiff : real ;
   j           : widthoftable ;
   powerof4    : posint ;

BEGIN
diff[p-1] := 0.0 ;
(* Call traprule to start a new row *)
rj := rtable[0] ;
traprule (false, rj, h, n) ;
powerof4 := 4 ;
FOR j := 0 TO p-1 DO
   BEGIN
   (* Update the differences and convergence ratios *)
   newdiff := rj - rtable[j] ;
   rtable[j] := rj ;
   IF abs(newdiff) > tolmin(rj) THEN
       cratio[j] := diff[j] / newdiff
   ELSE
       cratio[j] := 0.0 ;
   diff[j] := newdiff ;
   (* Compute the next extrapolated value *)
   rj := rj + newdiff / (powerof4 - 1) ;
   powerof4 := 4*powerof4
   END ;
rtable[p] := rj ;
IF trace THEN
   traceoftable (* write the new row to the output file *)
END (* rombstep *)
```

Notes

(i)   Procedure traprule given in section 9.1 is used to compute the trapezoidal estimate which is placed in rtable[0] .

(ii)   Before we compute a convergence ratio, we check that the difference
$T_{i,j} - T_{i-1,j}$ is non-zero to machine precision (using tolmin). If the difference is
zero we set the convergence ratio to 0.0.

(iii)   If the global variable trace = true, a procedure traceoftable is invoked to
print a trace. The code is

```
PROCEDURE traceoftable ;

    VAR
        j : widthoftable ;

    BEGIN
    write ('rt') ;
    (* Global variable p records the number of
        extrapolation columns *)
    FOR j := 0 TO p DO
        write (rtable[j]:14) ;
    IF p >= 2 THEN
        BEGIN
        writeln ;  write ('cr') ;
        FOR j := 0 TO p-2 DO
            write (cratio[j]:10, '     ') ;
        writeln
        END ;
    writeln
    END (* traceoftable *)
```

The output produced by traceoftable for the integral $\int_0^1 e^{\sin x}\, dx$ is of the form

```
rt 1.659888E+00
rt 1.637517E+00 1.630060E+00
rt 1.633212E+00 1.631776E+00 1.631891E+00
cr 5.20E+00

rt 1.632201E+00 1.631864E+00 1.631870E+00 1.631870E+00
cr 4.26E+00      1.96E+01

rt 1.631952E+00 1.631869E+00 1.631870E+00 1.631870E+00 1.631870
cr 4.06E+00      1.68E+01      7.69E+01
```

Here, traceoftable has been called five times. The rows of the Romberg table are
indicated by rt and the convergence ratios by cr (recall that three values in a
column are required before a convergence ratio can be calculated). We see that
the ratios are converging to 4^{j+1}, $j = 0, 1, 2, \ldots$, as the theory earlier in this
section would predict.
    To apply the cautious extrapolation test to the convergence ratios we have
the following procedure.

```
PROCEDURE cratiotest ( j           : widthoftable ;
                  VAR cratio       : rowvector ;
                  VAR convtest     : boolean ;
                  VAR extrpconst   : real ) ;
(* Applies the relaxed form of de Boor's cautious
    extrapolation test to decide if the extrapolated value
    in column j+1 should be accepted. If so, convtest is
```

```
returned as true, otherwise false. The parameter
extrpconst, which is a power of 4, is updated; it is used
in the main body of romberg to compute errbnd as
abs(diff[j])/(extrpconst-1) (if convtest is true) *)

CONST
   h2tol = 0.15 ;  (* tolerance for testing convergence
                       in trapezoidal column *)
   extol = 0.10 ;  (* tolerance for testing convergence
                       in extrapolation columns *)
VAR
   ratio : real ;

BEGIN
IF j = 0 THEN
   BEGIN  (* test for O(h^2) convergence in column 0 *)
   convtest := abs (cratio[0] - 4.0) < h2tol ;
   extrpconst := 4.0
   END
ELSE
   BEGIN
   ratio := cratio[j] ;
   IF ratio <> 0.0 THEN
      BEGIN
      convtest := ratio > extrpconst ;
      IF convtest THEN
         IF abs (ratio/4 - extrpconst)
                              < extol * extrpconst THEN
            extrpconst := 4*extrpconst
      END
   ELSE
      convtest := true
   END
END (* cratiotest *)
```

On exit, convtest = true if the convergence ratio for column j passes the relaxed form of the cautious extrapolation test; if it fails, convtest = false.

If convtest is returned as false when j = 0, then we have failed to detect $0\ (h^2)$ convergence in the trapezoidal estimates. In this case we apply the special test 9.15, using a **boolean** function.

```
FUNCTION trapconvtest ( cr0, lastcr0 : real ) :  boolean ;
(* Returns true if the trapezoidal estimates appear to be
   converging faster than O(h), otherwise false. On entry
   cr0 should contain the current value of cratio[0];
   lastcr0 should contain the previous value *)

CONST
   convtol = 0.1 ;  (* tolerance for testing if the
                       trapezoidal values are converging *)
   cr0tol  = 2.0 ;  (* threshold for acceptably high
                       order of convergence *)

BEGIN
IF abs(cr0) > smallreal THEN
   trapconvtest := (abs(cr0-lastcr0) < convtol*abs(cr0))
                   AND (cr0 > cr0tol)
ELSE
   trapconvtest := false
END (* trapconvtest *)
```

If trapconvtest is returned as false, this means we cannot find a reliable error bound for the trapezoidal estimate in rtable[0].

In the main body of romberg we shall initialise the parameter errbnd (estimated error bound) to the default value 1/smallreal; it will remain at this value until we detect that the order of convergence is higher than $O(h)$. If errbnd is still at the default value when maxrow is reached, then the Romberg process is terminated with eflag set to 2.

Robustness and Adaptability Checks

Before we write the code for the body of romberg, we must ensure as far as possible that the algorithm is robust and adaptable. Four questions will be considered.

(i)   How can we guard against the stepwidth becoming too small for the computer?

(ii)   How do rounding errors affect the computed value of the integral?

(iii)   If the first two estimates $T(b-a)$ and $T((b-a)/2)$ produced by traprule are equal, does this mean that the integrand is a straight line?

(iv)   If the estimates produced are zero, does this mean that the value of the integral is zero?

(i)   *Limiting the stepwidth*

If the stepwidth becomes too small relative to the precision of the computer, we are unable to distinguish between the integration points. From chapter 4 we know that the floating-point representations of two numbers x and $x+h$ can be distinguished provided that

$$|h| \geqslant \max (\text{tolmin}(x), \text{tolmin}(x+h))$$

Using tolmin, it is easy to write a function to compute the minimum safe stepwidth h such that

$$fl(x+h) \neq fl(x) \quad \text{for } x \in [a, b]$$

```
FUNCTION stepmin ( a, b : real ) :  real ;
(* Computes the smallest value which can be used safely
   as an integration stepwidth in the interval [a, b].
   Routine called from the library:  tolmin *)

   BEGIN
   IF abs(a) > abs(b) THEN
      stepmin := tolmin(a)
   ELSE
      stepmin := tolmin(b)
   END (* stepmin *)
```

Since we shall use this function again in a later integration routine, we include it in Group 0 of mathlib (library utility routines).

(ii)   *Rounding errors*

In our discussion of numerical integration so far we have largely ignored rounding errors. This is because they have no significant effect unless the required accuracy is close to machine precision. However, our usual terminating test as developed in chapter 4, which would now take the form

    errbnd < tol + tolmin(ivalue)

needs modification, because rounding errors arising in the trapezoidal computation seriously affect the Romberg process as machine precision is approached. We must cut off the iteration before this can happen, and we therefore use a modified test

$$\text{errbnd} < \text{tol} + \text{limacc} * \text{tolmin(ivalue)} \tag{9.16}$$

The constant limacc, which must be substantially greater than 1, limits the accuracy which can be obtained from procedure romberg (effectively we are using a computer of reduced precision). From numerical experiments we find that a suitable value for limacc is 100; this allows us to use the integrator with tol set to 0.0 and still obtain valid results.

(iii) *Testing for linearity*

If the first two estimates produced by traprule are equal, then diff[0] = 0.0. There are two possibilities: either $f(x)$ is a straight line (making the trapezoidal rule exact), or $f(x)$ happens to interpolate a straight line at the three points a, $(a + b)/2$ and b (for example, it may be zero at all three points). To test for linearity we use a boolean function.

```
FUNCTION linetest :  boolean ;
(* Returns true if the integrand f(x) appears to be a
   straight line ,to machine precision *)

   VAR
      step, incr, xr, fa, fb, fx : real ;
      i       : integer ;
      linear : boolean ;

   BEGIN
   fa := f(a) ;   fb := f(b) ;
   step := b - a ;   incr := fb - fa ;
   i := 0 ;
   REPEAT
      (* Use four "random" points in (a, b) to test for
         linearity *)
      i := i + 1 ;
      CASE i OF
         1: xr := 0.71420053 ;
         2: xr := 0.34662815 ;
         3: xr := 0.84375100 ;
         4: xr := 0.12633046
         END (* CASE *) ;
      fx := f(a + xr*step) ;
      linear := abs(fx-(fa+xr*incr)) < limacc * tolmin(fx)
   UNTIL NOT linear OR (i = 4) ;
   linetest := linear
   END (* linetest *)
```

(Since rounding errors can invalidate the test, we again take the precaution of using limacc.)

(iv) *Testing for zero integral*

There is one further situation that requires special attention. For an integral
such as

$$\int_{-1}^{1} x^3 \, dx = 0$$

the trapezoidal values (and hence the values in diff[0]) are zero on every row of
the Romberg table. The difficulty is that the value of the integral is actually
zero. We are unable to estimate the order of convergence, and so errbnd remains
at the default value 1/smallreal; unless we take some special action, romberg will
fail with eflag set to 2. We handle this by testing the integrand for antisymmetry
about the mid-point of the interval, using a boolean function similar to linetest.

```
FUNCTION oddtest :  boolean ;
(* Returns true if f(x) appears to be antisymmetric about
   the mid-point of [a, b] to machine precision *)

   VAR
      halfstep, mid, dx, fx : real ;
      i     : integer ;
      oddf : boolean ;

   BEGIN
   halfstep := (b-a)/2 ;
   mid := a + halfstep ;
   i := 0 ;
   REPEAT
      i := i + 1 ;
      CASE i OF
         1: dx := 0.62481392 * halfstep ;
         2: dx := 0.28345761 * halfstep ;
         3: dx := 0.47262394 * halfstep
         END (* CASE *) ;
      fx := f(mid+dx) ;
      oddf := abs(fx + f(mid-dx)) < limacc * tolmin(fx)
   UNTIL NOT oddf OR (i = 3) ;
   oddtest := oddf
   END (* oddtest *)
```

At last we can give the code for the main body of romberg.

```
BEGIN   (* body of PROCEDURE romberg *)
(* Initialisation *)
ivalue := 0.0 ;
converged := (a = b) OR oddtest ;
IF converged THEN
   errbnd := 0.0
ELSE

   BEGIN   (* Romberg integration *)
   errbnd := 1/smallreal ;
   smallh := 2 * stepmin(a, b) ;
   row := 1 ;   (* current row of the Romberg table *)
   p := 0 ;      (* number of extrapolation columns *)
   (* Call traprule to compute and store initial estimate *)
   traprule (true, rtable[0], h, n) ;
   IF trace THEN
      BEGIN
      writeln ;
      writeln ('*** trace in romberg ***') ;  writeln ;
      traceoftable  (* output first element *)
      END ;
```

```
REPEAT
    limith := abs(h) < smallh; (* limiting precision test *)
    IF NOT limith THEN
        BEGIN
        row := row + 1 ;
        IF p < maxcol THEN
            p := p + 1 ;
        (* Call rombstep to construct a new row with
           updated differences and convergence ratios *)
        rombstep (p, rtable, diff, cratio, h, n) ;
        ivalue := rtable[0] ;  (* trapezoidal estimate *)
        IF row = 2 THEN
            BEGIN
            (* insufficient values to calculate convergence
               ratios; test for linearity of f(x) *)
            IF abs(diff[0]) < tolmin(ivalue) THEN
                BEGIN
                converged := linetest ;
                IF converged THEN
                    errbnd := 0.0
                END
            END


        ELSE
            BEGIN (* row > 2 *)
            tola := tol + limacc * tolmin(ivalue) ;
            j := 0 ;
            REPEAT
                (* apply cautious extrapolation test *)
                cratiotest (j, cratio, convtest, extrpconst) ;
                IF convtest THEN
                    BEGIN  (* value in rtable[j+1] accepted *)
                    ivalue := rtable[j+1] ;
                    errbnd := abs(diff[j]) / (extrpconst-1) ;
                    converged := errbnd < tola ;
                    j := j + 1
                    END
            UNTIL (j = p-1)   (* last possible test on p-2 *)
                OR converged  (* value sufficiently accurate *)
                OR NOT convtest ; (* failed extrapolation
                                                      test *)
            IF NOT convtest AND (j < p-2) THEN
                p := j + 2 ;  (* restrict the number of columns
                                     while extrapolation invalid *)
            IF j = 0 THEN  (* O(h^2) convergence not detected
                                  in trapezoidal estimates *)
                IF (row > 3)
                AND trapconvtest (cratio[0], lastcr0) THEN
                    BEGIN  (* convergence faster than O(h) *)
                    errbnd := abs(diff[0]) ;
                    converged := errbnd < tola
                    END ;
            lastcr0 := cratio[0]
            END (* row > 2 *)
        END
UNTIL converged         (* sufficient accuracy attained *)
    OR limith           (* limiting precision reached   *)
    OR (row = maxrow) ; (* maximum number of rows used  *)

END (* Romberg integration *) ;

IF errbnd < tol THEN
    eflag := 0   (* full requested accuracy attained *)
ELSE
    IF errbnd < 1/smallreal THEN
        errormessage ('romberg ', 1,
            'required accuracy not attained', warning, eflag)
```

```
ELSE
    errormessage ('romberg ', 2,
        'no error bound has been found ', fatal, eflag)
END (* romberg *)
```

From the user's point of view there are three possible outcomes to a call of romberg (apart from runtime failures such as overflow in f(x), which we do not claim to have eliminated).

(i)   eflag = 0   Reliable results are returned in ivalue and errbnd, and the estimated error bound lies strictly within the specified tolerance tol. The computation has been entirely successful.

(ii)   eflag = 1   Results are returned for the integral and an estimated error bound, but the latter does not fall within tol. This may happen because the minimum stepwidth of the computer has been reached, or the condition 9.16 has been satisfied and the iteration terminated for the reason discussed there, or the maximum number of rows of the Romberg table have been used up (normally when the trapezoidal estimates have faster than $0(h)$ but not $0(h^2)$ convergence). Depending on the value returned in errbnd, the results may still be of use.

(iii) eflag = 2   The routine has been unable to determine a reliable error bound because the integrand is too badly behaved over the interval of integration. The parameter errbnd is returned with the default value 1/smallreal, and the 'result' in ivalue cannot be believed.

The user must be careful about interpreting the results produced by romberg or any similar routine. It is assumed that the integrand $f(x)$, as specified in the Pascal function f(x), is correct to machine precision. Any errors in the actual integrand will affect the accuracy that can be obtained from romberg.

Numerical Testing

As procedure romberg is more complicated than our earlier routines, it is more difficult to devise a set of test problems to exercise the code adequately. Space does not permit us to give full details of all our numerical testing; we shall concentrate on exercising the main paths through the cautious extrapolation method. We illustrate the process by using the trace option. The accuracy requested in all cases is six correct decimal places (tol = 0.5E − 6), and the number of evaluations of the integrand is given as a measure of efficiency.

Test Integral 1

$$\int_0^1 \frac{4}{1+x^2} \, dx = \pi$$

```
*** trace in romberg ***

rt 3.000000E+00
rt 3.100000E+00 3.133333E+00
rt 3.131176E+00 3.141569E+00 3.142118E+00
cr 3.21E+00

rt 3.138988E+00 3.141593E+00 3.141594E+00 3.141586E+00
cr 3.99E+00      3.45E+02

rt 3.140942E+00 3.141593E+00 3.141593E+00 3.141593E+00 3.141593
cr 4.00E+00      1.60E+02      3.65E+02


computed value of integral:   3.141592661143E+00
estimated error bound:        4.958870492011E-08
function evaluations:         19
```

The integrand $4/(1 + x^2)$ is very well-behaved, and we see that the convergence ratio for the trapezoidal column has already converged to 4.00 correct to three significant figures.

Test Integral 2

$$\int_0^1 (4x - 1)\,dx = 1$$

```
*** trace in romberg ***

rt 1.000000E+00
rt 1.000000E+00 1.000000E+00

computed value of integral:   1.000000000000E+00
estimated error bound:        0.000000000000E+00
function evaluations:         11
```

Function linetest has correctly detected that the integrand is a straight line.

Test Integral 3

$$\int_0^1 e^{-x} \sin 8\pi x\,dx = 0.02511152\ldots$$

```
*** trace in romberg ***

rt 0.000000E+00
rt 0.000000E+00 0.000000E+00
rt 0.000000E+00 0.000000E+00 0.000000E+00
cr 0.00E+00

rt 0.000000E+00 0.000000E+00 0.000000E+00 0.000000E+00
cr 0.00E+00      0.00E+00

rt 1.971525E-02 2.628700E-02 2.803946E-02 2.848454E-02
cr 0.00E+00      0.00E+00
            .
rt 2.380522E-02 2.516854E-02 2.509398E-02 2.504722E-02
cr 4.82E+00     -2.35E+01

rt 2.478747E-02 2.511489E-02 2.511131E-02 2.511159E-02
cr 4.16E+00      2.08E+01
```

```
rt 2.503067E-02 2.511173E-02 2.511152E-02 2.511152E-02
cr 4.04E+00      1.70E+01
```

```
computed value of integral:    2.511151968066E-02
estimated error bound:         2.107041503162E-07
function evaluations:          134
```

This shows that romberg, unlike the algorithm in section 9.1, is not deceived when the early trapezoidal values are all zero; function linetest correctly decides that the integrand is not a straight line.

Test Integral 4

$$\int_0^1 5x^4 \, dx = 1$$

```
*** trace in romberg ***

rt 2.500000E+00
rt 1.406250E+00 1.041667E+00
rt 1.103516E+00 1.002604E+00 1.000000E+00
cr 3.61E+00

rt 1.026001E+00 1.000163E+00 1.000000E+00 1.000000E+00
cr 3.91E+00     1.60E+01

rt 1.006508E+00 1.000010E+00 1.000000E+00 1.000000E+00 1.000000
cr 3.98E+00     1.60E+01      0.00E+00
```

```
computed value of integral:    1.000000000000E+00
estimated error bound:         0.000000000000E+00
function evaluations:          19
```

Here the values in the second extrapolation column are exact, and the convergence ratio is therefore zero (in general, the values in the jth extrapolation column are exact if the integrand is a polynomial of degree less than or equal to $2j + 1$). To handle such cases we set convtest to true in procedure cratiotest if the convergence ratio is zero to machine precision.

Test Integral 5

$$\int_0^1 \sqrt{(1 - x^2)} \, e^x \, dx = 1.24395050 \ldots$$

We list rows 13, 14 and 15 of the Romberg table.

```
*** trace in romberg ***

  .   .   .   .   .   .   .   .   .   .   .   .   .   .   .   .   .
  .   .   .   .   .   .   .   .   .   .   .   .   .   .   .   .

rt 1.243947E+00 1.243949E+00 1.243949E+00 1.243949E+00
cr 2.83E+00     2.83E+00

rt 1.243949E+00 1.243950E+00 1.243950E+00 1.243950E+00
cr 2.83E+00     2.83E+00
```

```
rt 1.243950E+00 1.243950E+00 1.243950E+00 1.243950E+00
cr 2.83E+00     2.83E+00

*** warning in romberg : 1   required accuracy not attained

computed value of integral:   1.243950120039E+00
estimated error bound:        6.976741895315E-07
function evaluations:         16387
```

Only the trapezoidal column is relevant. We have already remarked that for this integral the first derivative of the integrand is singular at $x = 1$ and the Euler-Maclaurin expansion is not valid. This is reflected in the results. The convergence ratio for the trapezoidal column appears to be converging to 2.83, which shows that the trapezoidal values do not have $0\,(h^2)$ convergence; procedure romberg therefore does not accept any extrapolated estimates. However, the convergence in the trapezoidal column is sufficiently rapid for romberg to be able to bound the error; maxrow is reached before the error bound falls within $0.5E - 6$, so romberg returns eflag = 1†. The user who wished to go further could partition the interval $[0, 1]$ into two subintervals and apply romberg to each part; since the difficulty arises at $x = 1$, a sensible partition would be $[0.0, 0.95] \cup [0.95, 1.0]$ (in general, however, it would be difficult for the user to determine such a partitioning).

The above set of test integrals is clearly far from completely exercising procedure romberg (for example, we have not tested the limiting precision paths, and we have used the interval $[0, 1]$ throughout). It is difficult and time-consuming but nevertheless necessary to exercise a library code to the fullest possible extent. Without software tools to assist in this we can never be completely confident that we have exercised all the paths.

Procedure romberg illustrates the problems of writing mathematical software designed for the general user. Starting from a sound mathematical method, we have added a number of special tests in an attempt to ensure that the algorithm is reliable and robust. These tests typically involve *ad hoc* constants (for example, h2tol = 0.15) which are chosen on the basis of a mixture of mathematical theory and numerical experience. This can have a dramatic effect on the performance of the routine; it may be regarded as part of the *art* of mathematical software.

9.4   ADAPTIVE SIMPSON'S RULE

Romberg integration is called a *whole-interval process* because all the entries in the Romberg table are estimates of the integral over the whole interval $[a, b]$ (obtained, of course, with different numbers of subdivisions and different orders of approximation). As we have seen in Test Integral 5, $\int_0^1 \sqrt{(1 - x^2)}\, e^x \, \mathrm{d}x$, this

---

†The computed value of the integral is actually correct to 6 decimal places, but romberg does not detect this because the error bound falls just outside the tolerance.

is not always a good approach. It may be more efficient to partition $[a, b]$ into
a number of subintervals of various sizes and compute the integral over each
subinterval separately. Where the integrand is changing rapidly (for example,
near $x = 1$ in figure 9.1), we would take a small interval such as $[0.95, 1.0]$ to
obtain an accurate result; elsewhere, we could obtain the same accuracy with a
much larger interval. Thus we concentrate our efforts (evaluations of $f(x)$) in
places where they are most needed.

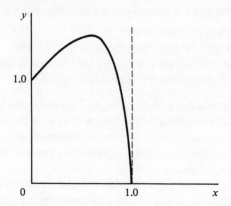

Figure 9.1   Graph of $y = \sqrt{(1 - x^2)}\, e^x$ over $[0, 1]$

It would be very inconvenient in general for the user to have to supply a suitable partitioning, so we look for a method of doing this automatically. The idea
of partitioning the interval $[a, b]$ automatically is the basis for a class of methods
called *adaptive integration schemes*.

To describe a simple adaptive scheme, suppose we have a rule $Q(J, h)$ which
approximates an integral over an interval J using a stepwidth h, and also a
formula $E(J, h)$ for estimating the truncation error $\int_J f(x)\, dx - Q(J, h)$ (for
example, Q could be any of the composite fixed-point rules developed in
chapter 8). We wish to estimate the integral $I = \int_a^b f(x)\, dx$ to an absolute error
tolerance tol. First we calculate an estimate of the integral

$$\overline{Q} = Q([a, b], \overline{h}) \approx I$$

and the corresponding estimate of truncation error

$$\overline{E} = E([a, b], \overline{h}) \approx I - \overline{Q}$$

for a suitable stepwidth $\overline{h}$ (in the composite trapezoidal rule, for example, $\overline{h}$
could be $(b - a)/2$). We accept $\overline{Q}$ as sufficiently accurate if $|\overline{E}| < $ tol. Otherwise
we partition the interval $[a, b]$ into two subintervals, $[a, x_m]$ and $[x_m, b]$,
where $x_m = (a + b)/2$. We then calculate $\overline{Q}_l = Q([a, x_m], \overline{h}/2)$ and
$\overline{Q}_r = Q([x_m, b], \overline{h}/2)$, together with the corresponding error estimates $\overline{E}_l$ and
$\overline{E}_r$. In order to accept the estimate $\overline{Q}_l + \overline{Q}_r$ for the integral I, we require that

$$|\overline{E}_l + \overline{E}_r| < \text{tol}$$

for which a sufficient condition is

$$(|\bar{E}_l| < \tfrac{1}{2}\,\text{tol})\quad \text{AND}\quad (|\bar{E}_r| < \tfrac{1}{2}\,\text{tol})$$

If either of these subordinate conditions is satisfied, we accept the estimate $\bar{Q}_l$ or $\bar{Q}_r$ for the corresponding subintegral. If either or both of the conditions are not satisfied, we halve the tolerance and apply the process again to the subintegral(s) in question.

In general, an adaptive integration scheme has three main components:

(i)   a fixed-point integration rule $Q(J, h)$;
(ii)   a formula $E(J, h)$ for estimating the error of $Q(J, h)$;
(iii)   a strategy for selecting the subinterval to subdivide at each stage and for deciding when the over-all accuracy requirement is satisfied.

We shall develop an adaptive scheme based on Simpson's rule and estimate the local error by an extrapolation method. Simpson's rule is chosen so that the coding of the algorithm as a library routine, procedure adsimp, can be kept comparatively simple (the use of other integration rules will be discussed briefly at the end of this section)†.

We consider a subinterval $[x_l, x_r] \subseteq [a, b]$ and suppose that we wish to compute the subintegral $I_S = \int_{x_l}^{x_r} f(x)\,\mathrm{d}x$ to a tolerance tol$_S$. From Simpson's rule with a stepwidth $h = (x_r - x_l)/2$ we have an estimate

$$S_1 = S([x_l, x_r], h) = \frac{h}{3}\ (f(x_l) + 4f(x_m) + f(x_r))$$

where $x_m = (x_l + x_r)/2$. As we shall see, the value of S_1 will be passed down from a higher level of recursion. What we must do at this stage is to bisect $[x_l, x_r]$ and evaluate Simpson's rule for a stepwidth $h/2$. This can be expressed as

$$S_{\frac{1}{2}} = S([x_l, x_r], h/2)$$
$$= S([x_l, x_m], h/2) + S([x_m, x_r], h/2) \tag{9.17}$$

That is, we apply the basic form of Simpson's rule to each half of the interval $[x_l, x_r]$. The only 'new' points occurring are at $(x_l + x_m)/2$ and $(x_m + x_r)/2$, so only two new function evaluations need be performed.

From the analysis in section 9.2 we know that, provided f is at least six times continuously differentiable on $[x_l, x_r]$, the error of the estimate S_1 is given by

$$I_S - S_1 = \bar{b}_4 h^4 + 0(h^6)$$

where $\bar{b}_4$ is a constant independent of h. Rewriting this for $h/2$ and eliminating the term in h^4, we obtain

†The adaptive scheme used in adsimp was originally described by W. M. Mckeeman and later improved by J. N. Lyness; we refer to it as the Mckeeman/Lyness algorithm.

$$I_S - S_{\frac{1}{2}} = \frac{S_{\frac{1}{2}} - S_1}{15} + 0(h^6)$$

This provides us with a formula for estimating the error in $S_{\frac{1}{2}}$:

$$E_{\frac{1}{2}} = E\left([x_l, x_r], h/2\right) = \frac{S_{\frac{1}{2}} - S_1}{15} \qquad (9.18)$$

Now we must consider our partitioning strategy. Having computed $S_{\frac{1}{2}}$ from equation 9.17 and the error estimate $E_{\frac{1}{2}}$ from equation 9.18, we accept $S_{\frac{1}{2}}$ as sufficiently accurate if $|E_{\frac{1}{2}}| < \text{tol}_S$. If this condition is not satisfied, we take the two subintervals $[x_l, x_m]$ and $[x_m, x_r]$ of the current interval and apply the whole process again to each half (supplying the value $S([x_l, x_m], h/2)$ or $S([x_m, x_r], h/2)$, which now becomes S_1, and a tolerance of $\frac{1}{2}\text{tol}_S$).

Function simp

The algorithm describe above can be coded conveniently if we use the facility of recursive function calls in Pascal. The depth of recursion is recorded for trace purposes in an integer variable depth, and the integrand is specified by a Pascal function f(x); both are global to the following function.

```
FUNCTION simp (
             xl, xm, xr : real ;  (* left, middle and right
                                     points of subinterval *)
         lastsubivalue : real ;  (* previous estimate of the
                                     subintegral by Simpson's
                                     rule with h=(xr-xl)/2 *)
                subtol : real    (* absolute error tolerance
                                     for the subintegral    *)
                        ) : real ;

VAR
    xlm, lsubivalue  : real ;   (* for left half-interval *)
    xmr, rsubivalue  : real ;   (* for right half-interval *)
    h, subivalue,
            suberrest : real ;

BEGIN
writeln ('depth', depth:2, '  computing the subintegral',
         ' from ', xl:10, ' to ', xr:10) ;
h := (xr - xl)/4 ;                   (* new stepwidth and   *)
xlm := xl + h ;  xmr := xm + h ; (* integration points *)
(* Compute a new estimate of the subintegral by two
   applications of Simpson's rule *)
lsubivalue := (h/3) * (f(xl) + 4*f(xlm) + f(xm)) ;
rsubivalue := (h/3) * (f(xm) + 4*f(xmr) + f(xr)) ;
subivalue := lsubivalue + rsubivalue ;
suberrest := (subivalue - lastsubivalue)/15 ;

IF abs(suberrest) < subtol THEN
    (* required accuracy attained on [xl, xr]; return
       an extrapolated estimate of the subintegral *)
    simp := subivalue + suberrest
ELSE
```

```
   BEGIN
   (* local accuracy not yet attained; recall simp to
      approximate the subintegral on each half-interval *)
   depth := depth + 1 ;
   simp := simp (xl, xlm, xm, lsubivalue, subtol/2)
               + simp (xm, xmr, xr, rsubivalue, subtol/2) ;
   depth := depth - 1
   END
END (* simp *)
```

Notes

(i)   The previous estimate of the subintegral $\int_{x_l}^{x_r} f(x) \, dx$ obtained from Simpson's rule with a stepwidth $h = (x_r - x_l)/2$, that is, the value S_1, must be supplied as a parameter lastsubivalue. To start off the recursion, a value will be calculated with $h = (b - a)/2$ before the first call of simp.

(ii)   If the accuracy condition is satisfied on the subinterval $[x_l, x_r]$, Richardson's extrapolation is applied and the estimate of the subintegral is returned in simp.

(iii)   We have not concerned ourselves at this stage with economising on function evaluations or with questions of robustness and adaptability. Later, with refinements to improve the code in these respects, function simp will become function simpson.

To show the recursive nature of the algorithm we consider the integral

$$\int_{-2.5}^{4.0} \frac{1}{1 + x^2} \, dx = 2.51610761 \ldots$$

The following program calls simp to compute the integral correct to 2 decimal places.

```
PROGRAM simpexample ( output ) ;

CONST
   a = -2.5 ;   b = 4.0 ;
   tol = 0.5E-2 ;

VAR
   mid, ivalue : real ;
   depth       : integer ;

FUNCTION f ( x : real ) :  real ;
   BEGIN
   f := 1 / (1 + sqr(x))
   END (* f *) ;

<FUNCTION simp> ;

BEGIN  (* main program *)
mid := (a+b)/2 ;
ivalue := ((b-a)/6) * (f(a) + 4*f(mid) + f(b)) ;
(* Call the recursive function simp *)
depth := 1 ;
ivalue := simp (a, mid, b, ivalue, tol) ;
writeln ;
writeln ('computed value of integral: ', ivalue) ;
END .
```

The output is

```
depth 1  computing the subintegral from -2.50E+00 to  4.00E+00
depth 2  computing the subintegral from -2.50E+00 to  7.50E-01
depth 3  computing the subintegral from -2.50E+00 to -8.75E-01
depth 3  computing the subintegral from -8.75E-01 to  7.50E-01
depth 4  computing the subintegral from -8.75E-01 to -6.25E-02
depth 4  computing the subintegral from -6.25E-02 to  7.50E-01
depth 2  computing the subintegral from  7.50E-01 to  4.00E+00

computed value of integral:   2.514985024592E+00
```

The error in the computed value is 0.00112. . ., confirming that the required accuracy has been attained. The trace output from simp shows the subinterval considered at each stage. This is best illustrated by the tree structure in figure 9.2.

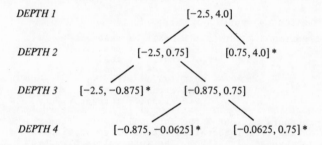

Figure 9.2   Subintervals considered by simp

The subintervals marked with an asterisk form the final partition of $[-2.5, 4.0]$ at which the required accuracy is attained.

To see why this partitioning arises, we consider the graph of the integrand $1/(1 + x^2)$. We use the library routine graf to plot the function values at the actual integration points used by simp.

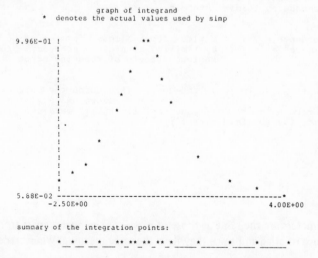

Figure 9.3   Integration points and function values used by simp

It can be seen that the integration points are clustered around $x = 0$ where the integrand $1/(1 + x^2)$ changes rapidly from a positive to a negative gradient. Further away from $x = 0$ the function is changing more slowly, and a larger interval can be used to obtain the same accuracy.

Procedure adsimp

We now consider how to make the algorithm robust and adaptable, and finally develop the code of the library routine adsimp. The heading and declarations are

```
PROCEDURE adsimp ( FUNCTION f(x:real)  : real ;
                             a, b       : real ;
                             tol        : real ;
                             trace      : boolean ;
                       VAR ivalue : real ;
                       VAR errbnd : real ;
                       VAR eflag  : integer ) ;

(* This procedure estimates the integral of f(x) from x=a to x=b
   using an adaptive integration scheme based on Simpson's rule
   (Mckeeman/Lyness algorithm). The interval is bisected
   recursively until sufficient accuracy can be obtained by two
   applications of Simpson's rule on each subinterval. The
   partitioning process is stopped if limiting precision of the
   computer is reached or the depth of recursion reaches a
   maximum of 25. The recursive behaviour of the algorithm can
   be exhibited by setting trace=true on entry.
   The error numbers for eflag are:
     =1  The required accuracy has not been attained -
         limiting precision of the computer reached  (warning)
     =2  The required accuracy has not been attained -
         maximum depth of recursion reached   (warning).
   Routines called from the library:
   errormessage, tolmin, stepmin      *)

CONST
    recdepth = 25; (* maximum depth allowed for recursion *)
    limacc  = 100; (* by limiting attainable accuracy, prevents
                      serious effects of rounding error   *)

VAR
    depth                 :  integer ;  (* records the current
                                           depth of recursion *)
    errorcondition   :  (noerror, limitprec, maxdepth) ;
    mid, fa, fb, fmid : real ;
```

Note

The parameters are the same as those of procedure romberg; apart from the significance of eflag and trace they have the same meaning. When trace = true, the depth of recursion and the subinterval currently being considered are written to the output file.

Apart from normal termination there are two possible error conditions in adsimp, one concerned with the limiting precision of the computer and the other with the maximum depth of recursion. If these conditions arise, they are detected in function simpson which forms the major part of adsimp. We use a global variable

    errorcondition: (noerror, limitprec, maxdepth)

to record the error condition.

The test for limiting precision on the subinterval [xl, xr] is

```
   (abs(h) < 2 * stepmin(xl, xr))
OR (abs(suberrest) < limacc * tolmin(subivalue))
```

where $h = (xr - xl)/4$. The first part of the test uses the mathlib utility routine stepmin, which was written originally for procedure romberg. If the condition is not true, we can safely call simpson to subdivide the interval further and the integration points will all be distinct. The second part of the test is intended to cut off the process, as in romberg, before growth of rounding error becomes a serious problem. From numerical experiments we find that a suitable value for the constant limacc is 100.

Restricting the depth of recursion is a necessary precaution to stop the process when further subdivision of an interval would not significantly improve the accuracy obtained. The method we are using is based theoretically on the assumption that the integrand is sufficiently differentiable. For integrals in which this is not the case, we have no guarantee that repeated partitioning of the interval will satisfy the accuracy requirement. The recursion depth cut-off, recdepth = 25, has been chosen on the basis of numerical experiments.

The variable errorcondition is initialised in adsimp to noerror and later set in simpson to the first error condition encountered, if any. The code of function simpson is now given.

```
FUNCTION simpson (
             xl, xm, xr : real ; (* left, middle and right
                                     points of subinterval *)
             fl, fm, fr : real ; (* function values at
                                     xl, xm and xr        *)
           lastsubivalue : real ; (* previous estimate of the
                                     subintegral by Simpson's
                                     rule with h=(xr-xl)/2 *)
               subtol : real     (* absolute error tolerance
                                     for the subintegral    *)
                      ) : real ;

(* The interval [xl, xr] is bisected and an estimate of the
   integral over [xl, xr] is computed by one application of
   Simpson's rule on each half-interval. If the estimated
   error is less than subtol, or an error condition
   (limiting precision or maximum depth of recursion) is
   detected, then Richardson's extrapolation is applied
   and the estimate of the integral is returned in simpson.
   Otherwise simpson is recalled to integrate over each
   half-interval separately *)
```

```
VAR
   xlm, flm, lsubivalue   : real ;   (* for left half *)
   xmr, fmr, rsubivalue   : real ;   (* for right half *)
   h, subivalue,
     suberrest, abserr     : real ;
   limitp                  : boolean ;

BEGIN
IF trace THEN
   writeln ('depth', depth:2, ' computing the subintegral'
            ,' from ', xl:10, ' to ', xr:10) ;
h := (xr - xl)/4 ;                  (* new stepwidth,       *)
xlm := xl + h ;  xmr := xm + h ;  (* integration points *)
flm := f(xlm) ;  fmr := f(xmr) ;  (* and function values *)
(* Compute a new estimate of the subintegral by two
   applications of Simpson's rule *)
lsubivalue := (h/3) * (fl + 4*flm + fm) ;
rsubivalue := (h/3) * (fm + 4*fmr + fr) ;
subivalue := lsubivalue + rsubivalue ;
suberrest := (subivalue - lastsubivalue)/15 ;
abserr := abs(suberrest) ;

IF abserr < subtol THEN
   BEGIN   (* required accuracy attained on [xl, xr];
               return an extrapolated estimate of the
               subintegral and update global error bound *)
   simpson := subivalue + suberrest ;
   errbnd := errbnd + abserr
   END
ELSE
   BEGIN   (* local accuracy not yet attained *)
   (* check for limiting precision *)
   limitp :=    (abs(h) < 2 * stepmin(xl, xr))
             OR (abserr < limacc * tolmin(subivalue)) ;
   IF NOT limitp AND (depth < recdepth) THEN
      BEGIN   (* recall simpson to approximate the integral
                  on each half of [xl, xr] *)
      depth := depth + 1 ;
      simpson := simpson (xl, xlm, xm, fl, flm, fm,
                                    lsubivalue, subtol/2)
                + simpson (xm, xmr, xr, fm, fmr, fr,
                                    rsubivalue, subtol/2) ;
      depth := depth - 1
      END
   ELSE
      BEGIN   (* an error condition has been detected *)
      IF errorcondition = noerror THEN
         (* record the error found *)
         IF limitp THEN
            errorcondition := limitprec
         ELSE
            errorcondition := maxdepth ;
      (* Return the best attainable estimate *)
      simpson := subivalue + suberrest ;
      errbnd := errbnd + abserr
      END
   END
END (* simpson *)
```

Notes

(i)   We have included the parameters fl, fm and fr in the parameter list o simpson so that the integrand need not be evaluated more than once at the integration point. There are exactly two new evaluations of f(x) each time simpson is called.

(ii)   The estimated error bound for the integral over the whole interval is recorded in a global variable errbnd, which is initialised to 0.0 before the first call of simpson. The final value of errbnd is simply the sum of the estimated error bounds over the final subintervals.

All that now remains for the body of adsimp is to initialise the global variables and parameters for the first call of simpson, call simpson, and issue a warning message if any error condition is detected.

```
BEGIN   (* body of PROCEDURE adsimp *)
IF trace THEN
   BEGIN
   writeln ;
   writeln ('*** trace in adsimp ***') ;   writeln
   END ;
errbnd := 0.0 ;   (* initialise estimated global error bound *)
errorcondition := noerror ;
IF abs(b-a)/2 > 2*stepmin(a, b) THEN
   BEGIN   (* estimate the integral using Simpson's rule
                on the whole interval [a, b] *)
   mid := (a + b)/2 ;
   fa := f(a) ;   fmid := f(mid) ;   fb := f(b) ;
   ivalue := ((b-a)/6) * (fa + 4*fmid + fb) ;
   (* Call recursive function simpson to refine estimate *)
   depth := 1 ;
   ivalue := simpson (a, mid, b, fa, fmid, fb, ivalue, tol) ;
   IF errbnd < tol THEN
      errorcondition := noerror ;   (* ignore any warnings if
                                         accuracy attained *)
   END
ELSE
   BEGIN   (* initial interval too small for computer *)
   errorcondition := limitprec ;
   ivalue := 0.0
   END ;
(* Inspect final error condition *)
CASE errorcondition OF
   noerror  :   eflag := 0 ;
   limitprec:   errormessage ('adsimp   ', 1,
        'limiting precision reached   ', warning, eflag) ;
   maxdepth :   errormessage ('adsimp   ', 2,
        'maximum depth of recursion   ', warning, eflag)
   END (* CASE *)

END (* adsimp *)
```

Notes

(i)   A check is made to see whether the estimated global error bound falls within the tolerance; if it does, any error condition which may have been reported by simpson is ignored.

(ii)   If the initial stepwidth $|b - a|/2$ is too small to be subdivided even once, adsimp returns eflag = 1 and ivalue = errbnd = 0.0.

Example 9.3

Use the library routines romberg and adsimp to estimate the integrals

(a) $\displaystyle\int_0^1 \frac{4}{1+x^2}\,dx$ (b) $\displaystyle\int_0^1 \sqrt{(1-x^2)}\,e^x\,dx$

correct to 2, 4, 6 and 8 decimal places and also to full machine accuracy. Compare the relative efficiency of the two integrators by tabulating the number of evaluations of the integrand.

A summary of results obtained on the Amdahl V/7 is

(a)

tol	romberg ivalue	errbnd	count	adsimp ivalue	errbnd	count
0.5E−2	3.14159250	0.3E−2	11	3.14211765	0.6E−3	4
0.5E−4	3.14159409	0.8E−5	11	3.14159409	0.4E−4	9
0.5E−6	3.14159266	0.5E−7	19	3.14159265	0.2E−6	33
0.5E−8	3.14159265	0.8E−9	35	3.14159265	0.1E−8	121
0.0	3.14159265	0.2E−12	131	3.14159265	0.1E−12	1157

(b)

tol	romberg ivalue	errbnd	count	adsimp ivalue	errbnd	count
0.5E−2	1.24237304	0.3E−2	67	1.23156795	0.2E−2	9
0.5E−4	1.24392604	0.4E−4	1027	1.24394782	0.2E−4	61
0.5E−6	1.24395012	0.7E−6	16387	1.24395050	0.1E−6	177
0.5E−8	(eflag returned as 1)			1.24395050	0.2E−8	533
0.0	(eflag returned as 1)			1.24395050	0.4E−13	11781

The integrand in (a) is arbitrarily differentiable everywhere on the real line. The conditions for Romberg integration are more than satisfied, and, as expected, romberg performs very efficiently; adsimp, based on a low-order integration rule, is comparatively less efficient for accuracies of more than 5 to 6 decimal places. The integrand in (b) has a derivative with a singularity at $x = 1$: romberg (with cautious extrapolation) can now only accept the trapezoidal estimates; adsimp is very much more efficient over the whole range of accuracies. □

Further Reading and Developments

A general account of the strategies which may be employed in adaptive integration schemes is given by Malcolm and Simpson (1975). These authors describe the strategy we have used in adsimp as a *local strategy*, because the decision whether to accept the estimate in a subinterval J or to subdivide J further is based solely on the error estimate E_J for that subinterval. It is generally more efficient to use a *global strategy* in which we consider the entire set $A = \{J\}$ of *active intervals* (the active intervals are those marked with an asterisk in figure 9.2). If the required accuracy

$$\sum_{J \in A} |E_J| < \text{tol}$$

has not been attained, we select the interval with the largest error estimate $|E_J|$ and subdivide that interval. To implement this, it is necessary to store information for all the members of the set A; by using a binary tree data structure (see exercises) it is possible to devise an efficient algorithm for this process.

The adaptive integrators used in practice are based on a variety of fixed-point integration rules, sometimes combined with Romberg integration (CADRE, mentioned in section 9.3, is a case in point). Two popular codes based on Simpson's rule are SQUANK and SQUAGE (see Malcolm and Simpson (1975)). An adaptive scheme using an eighth-order Newton–Cotes rule is described by Forsythe, Malcolm and Moler (1979).

In chapter 8 we saw that Gaussian rules are generally more efficient than Newton–Cotes rules, with the added advantage that they do not require the value of the integrand at the limits of integration. It may be wondered why we have not considered a Gaussian rule for our adaptive integration code. The reason is that it is more difficult to organise the partitioning process and estimate the truncation error efficiently. As we have seen, only two new evaluations of the integrand are required in each call of function simpson. The 2-point Gaussian rule, which has the same order of accuracy, uses two points *within* the interval $[x_l, x_r]$ (see section 8.3). In order to estimate the truncation error we could bisect $[x_l, x_r]$ and evaluate the Gaussian rule on each half-interval; however, because none of the four new points coincide with the two old ones, this clearly requires four new evaluations of the integrand.

Special formulae for estimating the truncation error of Gaussian rules can be devised which do not suffer from this problem. Further information can be found in the chapter on quadrature in the manual of the NAG library (Mark 8), where reference is made to a major library of numerical integration routines called QUADPACK.

9.5 TRANSFORMATION METHODS

Although we have tried to make our library routines romberg and adsimp as general as possible, there are still many integrals for which they are quite unsuitable unless the user first partitions the interval or transforms the integral. Some examples of integrals which give trouble to one or both of our integrators follow.

(i) $$\int_0^1 \sqrt{(1 - x^2)}\, e^x \, dx$$

As we have seen, romberg is very inefficient for this integral because the derivative of the integrand has a singularity at $x = 1$; adsimp performs quite well.

(ii) $$\int_0^1 \frac{\sin 64\pi x}{x} \, dx$$

romberg cannot compute an accurate value for this integral because the extrapolation process again is not valid. Also, the user must be careful in coding the Pascal function for the integrand, because the expression is not defined at $x = 0$ but only in the limit as $x \to 0$.

(iii) $\displaystyle\int_0^1 \ln x \cdot \cos 10\pi x \; \mathrm{d}x$

The integrand has a singularity at $x = 0$; neither of our integrators can handle this directly.

(iv) $\displaystyle\int_0^\infty \frac{1}{1+x^2} \; \mathrm{d}x$

One of the limits of integration is infinite. We could treat this by integrating over $[0, M]$, where M is chosen large enough so that the remainder $\int_M^\infty (1/(1+x^2)) \; \mathrm{d}x$ can be neglected, but this is rather unsatisfactory.

We can avoid many of these difficulties if we have access to adaptive integrators based on Gaussian rules, as in the NAG library and QUADPACK. As an alternative, we shall consider in this section how we might transform the interval of integration to extend the class of integrals which can be evaluated by romberg and adsimp. The transformation methods developed will be implemented in a library routine, procedure integral, which can call on the two basic integrators.

The HP Transformation

The transformation we shall describe is one used on Hewlett-Packard calculators, which compute definite integrals by Romberg integration (Kahan (1980)). We consider first the integral

$$I = \int_0^1 \sqrt{(1-x^2)} \; \mathrm{e}^x \; \mathrm{d}x$$

and attempt to transform it so that the derivative of the integrand is bounded in the interval of integration. We shall apply the cubic transformation

$$x = T(y) = 3y^2 - 2y^3$$

which has been chosen so that

$$T(0) = 0, \quad T(1) = 1$$

and

$$T'(0) = 0, \quad T'(1) = 0$$

On substituting $x = T(y)$ into the integral we obtain

$$I = \int_0^1 f_T(y) \; \mathrm{d}y$$

where

$$f_T(y) = \sqrt{\{1 - [T(y)]^2\}} \; \mathrm{e}^{T(y)} \, T'(y)$$

We note that since $T'(0) = 0$ and $T'(1) = 0$, the transformed integrand $f_T(y)$ is zero at the limits of integration $y = 0$ and $y = 1$.

We now investigate the behaviour of the derivative $f'_T(y)$ as $y \to 1$, since this is the cause of the trouble in the original integral. When $f_T(y)$ is differentiated by parts, since $T(1) = 1$ and $T'(1) = 0$, the only term in $f'_T(y)$ which may have a non-zero limit as $y \to 1$ is

$$\psi(y) = \frac{-T(y) \, T'(y)}{\sqrt{\{1 - [T(y)]^2\}}} \; e^{T(y)} \, T'(y)$$

$$= \frac{e^{T(y)} \, T(y) \, T'(y)}{\sqrt{[1 + T(y)]}} \times \frac{[-T'(y)]}{\sqrt{[1 - T(y)]}} \tag{9.19}$$

Now from the definition of $T(y)$ above we have

$$1 - T(y) = 2y^3 - 3y^2 + 1$$

and

$$-T'(y) = 6y^2 - 6y$$

On substituting into equation 9.19 we find

$$\psi(y) = \frac{e^{T(y)} \, T(y) \, T'(y)}{\sqrt{[1 + T(y)]}} \times \frac{6y \, (y - 1)}{\sqrt{[(y - 1)^2 \, (2y + 1)]}}$$

hence

$$\psi(1) = \frac{e \times 1 \times 0}{\sqrt{2}} \times \frac{6}{\sqrt{3}} = 0$$

Since $\lim_{y \to 1} f'_T(y) = \psi(1)$, we see that the derivative of the transformed integrand tends to zero as $y \to 1$. Thus the source of difficulty in the original integral has been removed.

We can apply the technique more generally to integrals of the form $I = \int_a^b f(x) \, dx$. We consider a mapping

$$T : [0, 1] \to [a, b]$$

defined by

$$x = T(y) = a_0 + a_1 y + a_2 y^2 + a_3 y^3$$

where the coefficients $a_0, \ldots, a_3$ are chosen so that

$$T(0) = a, \quad T(1) = b, \quad T'(0) = T'(1) = 0$$

It is easy to see that the polynomial which satisfies these requirements is

$$T(y) = a + 3(b - a) \, y^2 - 2(b - a) \, y^3 \tag{9.20}$$

The transformed integral is

$$I = \int_0^1 f_T(y)\,\mathrm{d}y$$

where

$$f_T(y) = f(T(y))\,T'(y)$$
$$= f(T(y)) \times 6\,(b-a)\,(y - y^2) \qquad (9.21)$$

By an extension of our earlier analysis we can show that the transformation will remove the singularity in $f'(x)$ at $x = a$ or $x = b$ or both whenever $f(x)$ has one of the forms

$$f(x) = g(x)\,(x-a)^p\,(b-x)^q$$

or

$$f(x) = g(x)\,|x - c|^p \qquad (c = a \text{ or } b)$$

provided that $0 < p$, $q \leqslant \frac{1}{2}$ and $g(x)$ is differentiable at the appropriate point a or b. (The second form reduces to our earlier example if we set $g(x) = e^x \sqrt{(1+x)}$, $p = \frac{1}{2}$ and $c = 1$).

We implement the HP transformation in two subprograms. The first simply computes the coefficients of the cubic polynomial given by equation 9.20; for this a and b are required, and it is assumed that the values are available in global variables a and b.

```
PROCEDURE setupcoeffs ;
(* Computes the coefficients of the transformation
   polynomial T(y). All the variables used are global *)

   BEGIN
   d  := 6*(b-a) ;
   a0 :=   a ;
   a2 :=   d/2 ;
   a3 := -d/3
   END (* setupcoeffs *)
```

If the integrand $f(x)$ is defined by a Pascal function f(x), we evaluate the transformed integrand $f_T(y)$ according to equation 9.21. This is done as follows†.

```
FUNCTION ftrans ( y : real ) :  real ;
(* Evaluates the transformed integrand fT(y) = f(T(y))*T'(y).
   The coefficients a0, a2, a3 and d are available globally
   after one call of setupcoeffs *)

   VAR
     x, ysq : real ;

   BEGIN
   IF (y = 0.0) OR (y = 1.0) THEN
      ftrans := 0.0  (* transformed integrand assumed to be
                        zero at the limits of integration *)
```

† We shall use ftrans in our final integration routine, procedure integral. It will then be necessary to add the function f as a parameter to ftrans.

```
ELSE
   BEGIN
   ysq := sqr(y) ;
   x  := a0 + ysq*(a2 + y*a3) ;
   ftrans := f(x) * d * (y-ysq)
   END
END (* ftrans *)
```

To estimate the integral $\int_0^1 f_T(y) \, dy = \int_a^b f(x) \, dx$, we first invoke setupcoeffs; we then call romberg or adsimp with the parameter f set to ftrans.

Example 9.4

Compare the performance of romberg and adsimp on the integral

$$\int_0^1 \sqrt{(1 - x^2)} \, e^x dx$$

for accuracies of 2 to 8 decimal places with and without the HP transformation.

The results without the transformation have been tabulated in example 9.3(b). These, and a corresponding set of results with the transformation, are compared in the following graph.

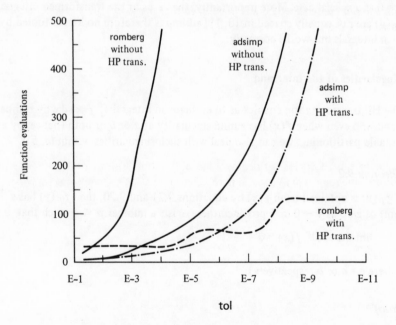

Figure 9.4   Performance of integrators on $\int_0^1 \sqrt{(1 - x^2)} \, e^x \, dx$ with and without the HP transformation

We also give part of the trace of romberg for the transformed integral.

```
*** trace in romberg ***

rt 0.000000E+00
rt 1.070876E+00 1.427835E+00
rt 1.211200E+00 1.257974E+00 1.246650E+00
cr 7.63E+00

rt 1.236061E+00 1.244349E+00 1.243440E+00 1.243389E+00
cr 5.64E+00     1.25E+01

rt 1.241993E+00 1.243970E+00 1.243945E+00 1.243953E+00
cr 4.19E+00     3.60E+01    -6.36E+00

rt 1.243462E+00 1.243952E+00 1.243950E+00 1.243951E+00
cr 4.04E+00     2.06E+01     8.88E+01

rt 1.243828E+00 1.243951E+00 1.243951E+00 1.243951E+00
cr 4.01E+00     1.71E+01     7.42E+01     1.81E+02
```

This should be compared with the trace for test integral 5 in section 9.3. The convergence ratios, denoted by cr, show that the columns are now all converging in a satisfactory manner. □

The HP transformation has other uses. From equation 9.21 it follows that $f_T(0) = f_T(1) = 0$ provided $f(x)$ does not have a singularity at a or b. Consequently, in ftrans the function f is not invoked at the limits of integration; ftrans simply returns 0.0. For an integral such as $\int_0^1 (1/x) \sin 64\pi x \, dx$, this relieves the user of the inconvenience of having to define the function value at $x = 0$ as a special case. More importantly, the zeros of the transformed integrand $f_T(y)$ are not equally spaced in $[0, 1]$; adsimp is therefore no longer fooled by such integrals but works correctly.

Singularities of the Integrand

The HP transformation enables us to evaluate an integral $\int_a^b f(x) \, dx$ by romberg or adsimp even when $f(x)$ has a mild singularity at a or b or both (hence, by a suitable partitioning, we can also deal with such singularities within (a, b)).

Theorem 9.2

If $f_T(y)$ is obtained from $f(x)$ by equations 9.21 and 9.20, then $f_T(y)$ has a limit of zero as $y \to 0$ or 1 provided there exists a number $p \le \frac{1}{2}$ such that

$$\lim_{x \to c} |x - c|^p f(x) = 0$$

(where $c = a$ or b respectively).

Proof

From equation 9.21 we have

$$f_T(y) = f(T(y)) \times 6(b - a) y (1 - y)$$

Provided $x \neq a$ this can be written as

$$f_T(y) = (x - a)^p f(x) \frac{6(b - a)y(1 - y)}{(T(y) - a)^p}$$

for any p. Substituting for $T(y) - a$ from equation 9.20, we obtain

$$f_T(y) = (x - a)^p f(x) \frac{6(b - a)y(1 - y)}{(b - a)^p y^{2p}(3 - 2y)^p}$$

It follows that if $\lim\limits_{x \to a} (x - a)^p f(x) = 0$ and $p \leqslant \frac{1}{2}$ then

$$\lim_{y \to 0} f_T(y) = 0$$

A similar proof can be given for $\lim\limits_{y \to 1} f_T(y)$. $\square$

The theorem shows that we can safely apply ftrans (which sets $f_T(0) = f_T(1) = 0$) to an integrand $f(x)$ with a singularity at one or both of the limits of integration provided $|f(x)|$ 'tends to infinity' more slowly than $K/\sqrt{x}$ for some constant K. For example, we can compute integrals of the type

$$\int_0^1 \left(\frac{1}{x^p} + \frac{1}{(1 - x)^q} \right) g(x) \, dx, \qquad p, q < \tfrac{1}{2}$$

where $g(x)$ is bounded on $[0, 1]$.

Example 9.5

Estimate the integral

$$\int_0^1 \ln x \cdot \cos 10\pi x \, dx$$

correct to 6 decimal places using romberg and adsimp.

It is known from analysis that $\lim\limits_{x \to 0} x^p \ln x = 0$ for any $p > 0$. The conditions of theorem 9.2 are satisfied (with $c = 0$), and we can apply ftrans followed by either of our integrators to compute the integral. The results are

```
***romberg***
computed value of integral:   -4.898866228433E-02
estimated error bound:        4.256826880277E-07
function evaluations:         8195


***adsimp***
computed value of integral:   -4.898881841790E-02
estimated error bound:        1.755458384744E-07
function evaluations:          459
```

Both integrators successfully estimate the integral, but because the transformed integrand is not a smooth function adsimp is considerably more efficient than romberg. $\square$

Introduction to Numerical Computation in Pascal

Infinite Limits of Integration

The semi-infinite intervals $[a, \infty), (-\infty, b]$ and the infinite interval $(-\infty, \infty)$ can be mapped onto $[0, 1]$ by a suitable transformation $x = V(z)$. For example, the substitution

$$x = a + (1 - z)/z$$

transforms the integral $\int_a^\infty f(x)\, dx$ to $\int_0^1 f_V(z)\, dz$, where

$$f_V(z) = f\left(a + \frac{1-z}{z}\right)\frac{1}{z^2}$$

We shall provide a set of three such transformations in one subprogram. First, we define a new library data type to distinguish the cases

```
intertype = (AtoB, AtoINF, MINFtoB, MINFtoINF)
```

The following function tests a global variable interval of type intertype and makes the corresponding transformation of the integrand; the variable save is also global and contains the value a or b as appropriate.

```
FUNCTION finf ( z : real ) :   real ;
(* Evaluates the transformed integrand fV(z) = f(V(z))*V'(z),
   where V is a mapping from [0, 1] to [a, inf), (-inf, b] or
   (-inf, inf). The global variable save contains the
   relevant value a or b *)

   BEGIN
   IF z = 0.0 THEN
      finf := 0.0   (* transformed integrand assumed to be zero
                       at limit corresponding to inf or -inf *)
   ELSE
      CASE interval OF
         AtoINF   :  finf := f(save + (1-z)/z) / sqr(z) ;
         MINFtoB  :  finf := f(save + (z-1)/z) / sqr(z) ;
         MINFtoINF:  finf := (f((1-z)/z) + f((z-1)/z)) /sqr(z)
         END (* CASE *)
   END (* finf *)
```

It is straightforward to show that under the mapping $V^{-1}: [a, \infty) \rightarrow [0, 1]$ the transformed integrand $f_V(z)$ is zero at $z = 0$ and finite at $z = 1$ if $f(a)$ is finite and $\lim_{x \to \infty} x^q f(x) = 0$ for some $q \geqslant 2$. For example, the integral $\int_0^\infty [1/(1 + x^3)]\, dx$ can be treated in this way. If now the HP transformation is applied from $[0, 1]$ to $[0, 1]$, it can be shown that the final transformed integrand is zero at both limits of integration under the (slightly weaker) conditions: $\lim_{x \to a} (x - a)^p f(x) = 0$ for some $p \leqslant \frac{1}{2}$ and $\lim_{x \to \infty} x^q f(x) = 0$ for some $q \geqslant 3/2$. This would allow, for example, the integral $\int_0^\infty [1/(1 + x^2)]\, dx$ to be treated. Similar sufficient conditions can be derived for integrals over $(-\infty, b]$ and $(-\infty, \infty)$.

Procedure integral, to be described next, permits the user to employ the function finf for semi-infinite or infinite intervals with or without ftrans; this allows the results to be compared for consistency. The same remark applies to finite intervals. When an integral is transformed by the substitution $x = T(y)$, the integrand f is evaluated at the points $x_i = T(y_i)$, where $\{y_i\}$ is the set of points selected by the integration rule. The sets $\{x_i\}$ and $\{y_i\}$ are quite different, and this enables the careful user to make consistency checks.

Procedure integral

We have seen even from our limited numerical testing that there is a place for both integrators. If the integrand is differentiable to a high order everywhere in the interval of integration and high accuracy is required, then romberg can be much more efficient than adsimp; in most other circumstances it is better to use adsimp. The user who wishes to compute integrals in practice must experiment with different integrators to find which is most suitable. To make this process more convenient we provide a library routine, procedure integral, through which romberg and adsimp can both be accessed (the integrators have been designed with compatible parameter lists). To enable the user to select the integrator, we define a new (and final) library data type

```
selectint = (rombergint, adsimpint)
```

Procedure integral also incorporates the transformation methods which we have developed to extend the applicability of the integrators.

```
PROCEDURE integral ( FUNCTION f(x:real)   : real ;
                              a, b, tol    : real ;
                              routine      : selectint ;
                              interval     : intertype ;
                              transform    : boolean ;
                     VAR ivalue, errbnd    : real ;
                     VAR count, eflag      : integer ) ;

(* This procedure estimates the integral of f(x) between the
   limits a and b, either or both of which may be infinite.
   The user selects the integrator to be used by assigning one
   of the following values to the parameter routine:
      rombergint - Romberg integration (PROCEDURE romberg)
      adsimpint  - adaptive Simpson's rule (PROCEDURE adsimp).
   The type of interval is specified by the value of the
   parameter interval:
      AtoB       - the integral is over a finite interval [a, b].
      AtoINF     - the integral is over a semi-infinite interval
                   [a, inf). It is transformed to one over [0, 1]
                   with integrand f(a + (1-z)/z) / sqr(z).
                   The parameter b is not used.
      MINFtoB    - the integral is over a semi-infinite interval
                   (-inf, b]. It is transformed to one over [0, 1]
                   with integrand f(b + (z-1)/z) / sqr(z).
                   The parameter a is not used.
      MINFtoINF- the integral is over the infinite interval
                   (-inf, inf). It is transformed to one over [0, 1]
                   with integrand (f((1-z)/z) + f((z-1)/z)) /sqr(z).
                   The parameters a and b are not used.
   If the parameter transform = true, the integral is transformed
   to one over [0, 1] by the "HP transformation" x = T(y).
   This option should normally be used in any of the following
   cases:
      (1) the integrand is defined only in the limit at a or b
          (f(x) need not be specified at such points);
```

```
    (2) the integrand has a mild singularity at a or b;
    (3) the derivative of the integrand is singular at a or b;
    (4) one or both of the limits of integration is infinite;
    (5) a consistency check is required between the results
        produced with and without the transformation (the actual
        integration points used are different).
For sufficient conditions under which the transformation will
produce valid results in cases (2), (3) and (4), see the
summary of library contents 4.2.
The parameter count returns the number of evaluations of the
integrand f. The other parameters are the same as in romberg
and adsimp. The error numbers for eflag are:
    =1  corresponds to eflag=1 in routine selected by routine
    =2   .  .  .  .     eflag=2  .  .  .  .  .  .  .  .
Routines called from the library:
either romberg (errormessage, tolmin, stepmin)
or      adsimp (errormessage, tolmin, stepmin)  *)


CONST
   trace = false ;  (* the trace facility is not available *)

VAR
   a0, a2, a3, d, save : real ;


<PROCEDURE setupcoeffs> ;


<FUNCTION ftrans> ;


<FUNCTION finf> ;


FUNCTION fint ( x : real ) :  real ;
(* Evaluates the final transformed integrand at a point x in
   [0,1] or the unmodified integrand at a point x in [a, b] *)


   BEGIN
   count := count + 1 ;  (* number of evaluations of f *)
   CASE interval OF
      AtoB      : IF transform THEN
                     fint := ftrans(f, x)
                  ELSE
                     fint := f(x) ;
      AtoINF,
      MINFtoB,
      MINFtoINF : IF transform THEN
                     fint := ftrans(finf, x)
                  ELSE
                     fint := finf(x)
      END (* CASE *)
   END (* fint *) ;


BEGIN  (* body of PROCEDURE integral *)
count := 0 ;
IF interval <> AtoB THEN
   BEGIN
   CASE interval OF
      AtoINF  :  save := a ;  (* save the finite limit *)
      MINFtoB :  save := b ;  (* for use in finf        *)
      MINFtoINF:
      END (* CASE *) ;
   a := 0.0 ;  b := 1.0 ;  (* change limits of integration *)
   count := -1       (* f not invoked at one of the limits *)
   END ;
```

```
IF transform THEN
   BEGIN
   setupcoeffs ;
   IF interval = AtoB THEN
       BEGIN
       a := 0.0 ;   b := 1.0
       END ;
   count := -2        (* f not invoked at either limit *)
   END ;

CASE routine OF
   rombergint :   romberg (fint, a, b, tol, trace,
                                     ivalue, errbnd, eflag) ;
   adsimpint  :   adsimp (fint, a, b, tol, trace,
                                    ivalue, errbnd, eflag)
   END (* CASE *) ;
IF interval = MINFtoINF THEN
   count := 2*count  (* two evaluations of f(x) at
                          each call of fint *)
END (* integral *)
```

Notes

(i)   The actual integrand supplied to either romberg or adsimp is selected in the function fint. For example, when interval = AtoINF and transform = true, fint takes the value of ftrans(finf, x). In our earlier terminology this means that two transformations are being applied, V followed by T. If transform = false, then depending on the value of interval either finf(x) or f(x) is used directly.

(ii)   At the end of this chain of transformations the user-supplied function f(x) is evaluated. The parameter count returns the number of evaluations of f(x) during the integration process. This is to assist the user who may wish to experiment with various combinations of integrator and transformation.

Example 9.6

Use romberg and adsimp with the HP transformation to estimate the integral

$$\int_0^\infty \frac{1}{1+x^2}\ dx \quad (= \pi/2)$$

correct to 6 decimal places.

   We shall not give an example program at this stage but merely indicate how procedure integral could be used to solve the problem. Only integral and f(x), together with the relevant data types and variables, are declared in the main program. A typical calling sequence would be

```
integral (f, a, b, tol, adsimpint, AtoINF, true,
                  ivalue, errbnd, count, eflag)
```

The variables with values supplied by the user are

   a   : 0.0
   b   : undefined

```
tol    : 0.5E–6
eflag  : 0
```

The results are

```
***romberg***
computed value of integral:    1.570796325827E+00
estimated error bound:         1.417792349893E-07
function evaluations:             67
```

```
***adsimp***
computed value of integral:    1.570796329695E+00
estimated error bound:         1.676088735206E-07
function evaluations:             95
```

This shows that when the integral is transformed both integrators are successful, as theory would lead us to expect, and a comparatively small number of evaluations of the integrand is required. For an integrand such as $1/(1 + x^2)$ this hardly matters, but in more realistic cases it can be of crucial importance. □

PROJECTS

9.1   Show that the integral

$$I = \int_0^1 \left[\frac{\sqrt{u}}{u - 1} - \frac{1}{\ln u} \right] du$$

can be transformed to

$$\int_0^1 \left[\frac{2w^2}{(w - 1)(w + 1)} - \frac{w}{\ln w} \right] dw$$

by the transformation $u = w^2$ (Kahan (1980)).

Write a program which uses procedure integral to estimate the value of I with and without this transformation. Compare the performance of romberg and adsimp over a range of accuracies (the exact value of I is 0.0364899740...).

Use the library routine graf to plot the actual integration points used by each integrator; hence explain why romberg is more efficient on the transformed integral. Comment on which form of the integral is more suitable for adsimp.

By considering the effects of rounding errors, explain why the transformed integral should not be expressed in the form

$$\int_0^1 \left[\frac{2w^2}{(w^2 - 1)} - \frac{w}{\ln w} \right] dw$$

9.2   Investigate the possibility of computing the following integrals by the routines available in mathlib. Consider which transformations are desirable or necessary, and decide on theoretical grounds whether the results will be valid. Also consider whether any further preparation of the integral is appropriate in each case.

$$\int_{-\infty}^{\infty} e^{-x^2}\, dx \qquad\qquad \int_{-\infty}^{\infty} \frac{\cos x}{(1+x^2)}\, dx$$

$$\int_{0}^{\infty} \frac{e^{-x}}{x}\, \sin x\, dx \qquad\qquad \int_{-\infty}^{0} \frac{\sin x}{x\,|x^3 + 8\,|^{1/3}}\, dx$$

$$\int_{0}^{\infty} \frac{1}{(\sqrt{x})\,(1+x)}\, dx \qquad\qquad \int_{0}^{1} \sqrt{(1-x^2)}\, \ell n\, x\, dx$$

$$\int_{0}^{1} \frac{\ell n\, x}{x^2 - 1}\, dx \qquad\qquad \int_{0}^{1} (\ell n\, \sin \pi x\, -\, \ell n\, x)\, dx$$

$$\int_{-1}^{1} |\,2x^3 - 3x^2 + x\,|^{1/2}\, e^{\tan x}\, dx \qquad\qquad \int_{0}^{4} \sum_{n=0}^{4} \frac{\cos \pi\, (x - \frac{1}{2})}{x - n}\, dx$$

In cases where you believe that you can compute valid results, investigate the performance of romberg and adsimp over a range of accuracies. Where possible display the comparative performance of the two routines as a set of cost curves.

9.3 An alternative way of selecting the next subinterval to subdivide in an adaptive integration scheme is by the global strategy suggested by Malcolm and Simpson (see discussion in section 9.4). Write an automatic integrator which uses this strategy in conjunction with Simpson's rule, and compare its performance with that of adsimp and romberg on a set of test integrals.

To implement the global strategy you will need to use arrays to store the following information for each subinterval.

subinterval data

xl	:	Left end-point of the subinterval
h	:	width of the subinterval
fl, flm	:	values of the integrand at xl, xl + h/4,
fm, fmr, fr:		xl + h/2, xl + 3h/4, xl + h respectively
suberrest	:	estimate of the local error calculated by extrapolation (this can be obtained in terms of the five function values above)

At each stage of the partitioning process the strategy is to bisect the subinterval with the largest value of abs(suberrest). The process is continued until the sum of abs(suberrest) over all the active subintervals is less than tol, or limiting precision is reached.

To improve the efficiency of the search procedure, Malcolm and Simpson suggest that pointers to the active subinterval data should be arranged in a partially ordered tree

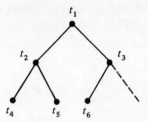

Figure 9.5   Pointers to the subinterval data structured as a tree

Node t_i contains the index of the array elements where the data for one of the active subintervals is stored. The nodes are ordered so that the value of abs(suberrest) associated with t_i is greater than the corresponding values for t_{2i} and t_{2i+1}. With this ordering, the next subinterval to be selected is always at the root of the tree. When this subinterval is bisected, the data associated with t_1 is overwritten by the data for the new half-interval with the larger value of abs(suberrest). The data for the other half-interval is stored in the next available location in the set of arrays. A new node t_n associated with this subinterval is placed at the bottom of the tree ($n = 7$ in the example above). The tree must then be reordered: first t_n is moved up the tree (in general t_i is moved up if the associated value of abs(suberrest) is greater than that for the parent of t_i); then t_1 is moved down in a similar manner.

When the integration process has converged, the value of the integral can be estimated by applying Simpson's rule twice on each active subinterval (using the values of the integrand saved for this purpose). It is possible to improve the estimate by Richardson's extrapolation.

Appendix A: Predeclared Mathematical Functions in Pascal

```
FUNCTION abs ( x : any arithmetic type ) :  same type ;
Returns the absolute value of x.


FUNCTION arctan ( x : real ) :  real ;
Computes the inverse tangent (in radians) of x.


FUNCTION cos ( x : real ) :  real ;
Computes the cosine of the angle x (considered to be
expressed in radians).


FUNCTION exp ( x : real ) :  real ;
Computes e^x.


FUNCTION ln ( x : real ) :  real ;
Computes the logarithm to base e of x.


FUNCTION odd ( i : integer ) :  boolean ;
Returns true iff i is odd.


FUNCTION round ( x : real ) :  integer ;
Returns x rounded to the nearest integer.


FUNCTION sin ( x : real ) :  real ;
Computes the sine of the angle x (considered to be
expressed in radians).


FUNCTION sqr ( x : any arithmetic type ) :  same type ;
Returns the square of x.


FUNCTION sqrt ( x : real ) :  real ;
Computes the square root of x.


FUNCTION trunc ( x : real ) :  integer ;
Returns x truncated to its integral part.
```

Appendix B: Summary of the mathlib Library

```
(*$E+ : compiler directive *)
PROGRAM mathlib ( input, output ) ;

(*            *******************
              The mathlib library
              *******************

   written by
           P. M. Dew  and  K. R. James,
     Department of Computer Studies, University of Leeds.

   Latest amendment:  16 March 1982.

   List of current routines:
     residu, errormessage, tolmin, mult, inprod, longinprod,
     stepmin;
     sign, powerr, poweri, graf, polsum, polval, roundt;
     polnewt, bisect, bisecant, bisrat;
     lufac, lusolv, linsolv, congrad;
     simpint, gaussint, romberg, adsimp, integral.

   Reference: "Introduction to Numerical Computation in Pascal"
        by P. M. Dew and K. R. James,  Macmillan Press, 1982. *)


(* ***************************
   External FORTRAN subroutine
   ***************************

   An external FORTRAN subroutine RESIDU is used in the linear
   algebra section of the library. This routine is to compute
   accurate residuals of a system of linear equations, and as
   this requires double precision arithmetic the routine cannot
   easily be written in Pascal. As far as the mathlib library is
   concerned, residu is a procedure in an external file which
   must be loaded with mathlib when the library is to be used.
   If this is not possible or not desired, then the library can
   be used with the following routines deleted:
        PROCEDURE residu,  PROCEDURE linsolv.
   A full specification of residu is given in the Summary of
   library contents below (Group 0: library utility routines).
   See also section 7.3 of the book.                            *)


(* ***************************
   Machine-dependent constants
   ***************************

   The library uses the following machine-dependent constants.
   The values given below apply to the Amdahl 470 using
   REAL*8; they should be modified for other computers. For
   further information see sections 2.2 and 4.3 of the book.   *)
```

```
CONST
   smallreal = 1.0E-74 ;   (* 100 times the smallest positive
                               real number *)
   rprec4    = 0.9E-15 ;   (* 4 times the relative precision
                               of the computer *)
   smallog   = -170.0  ;   (* ln(smallreal), used in tests to
                               avoid underflow failure *)
```

```
(* ********************************
   Definition of library data types
   ******************************** *)
```

```
   Because of limitations on the Pascal implementation, arrays
   in the library are of fixed size (see sections 2.2 and 2.4) *)
```

```
   upbnd = 100 ;    (* constants defining upper bounds  *)
   hibnd = 1000 ;   (* for the library array data types *)
```

```
TYPE
   (* To represent real and integer vectors:                 *)
   range   = 1 .. upbnd ;   (* range of the array subscript *)
   rvector = ARRAY [range] OF real ;
   ivector = ARRAY [range] OF integer ;

   (* To represent square matrices of real numbers:          *)
   matrix  = ARRAY [range, range] OF real ;

   (* To represent the coefficients of a polynomial:         *)
   index = 0 .. upbnd ;      (* range of the array subscript *)
   coeff = ARRAY [index] OF real ;

   (* For transmitting error messages:                       *)
   name    = PACKED ARRAY [1..8] OF char ;
   message = PACKED ARRAY [1..30] OF char ;
   severity = (fatal, warning) ;

   (* To represent long vectors in sparse linear systems:    *)
   longrange  = 1 .. hibnd ; (* range of the array subscript *)
   longvector = ARRAY [longrange] OF real ;

   (* To denote the set of positive integers:             *)
   posint = 1 .. maxint ;

   (* For selecting the Gaussian rule in PROCEDURE gaussint: *)
   gaussrules = 1 .. 4 ;

   (* For selecting the integrator and specifying the type
      of interval in PROCEDURE integral:                     *)
   selectint = (rombergint, adsimpint) ;
   intertype = (AtoB, AtoINF, MINFtoB, MINFtoINF) ;
```

```
(* ***************************
   Error messages and warnings
   *************************** *)
```

The last parameter in many of the library routines is eflag,
an error indicator. On entry, eflag should be set to 0 or -1
for the following actions:
```
   = 0  a message summarising fatal errors or warning of
        nonstandard terminating conditions is written to the
        output file via the library routine errormessage;
   =-1  no error messages or warnings are output.
```

On exit, eflag contains the value 0 if the computation has
been successful; otherwise, eflag contains the error number
as detailed in the specification of the library routine.

Present exceptions are the functions powerr and poweri, which
do not use the parameter eflag and write all error messages
to the output file.   *)


```
(* ***************************
   Summary of library contents
   ***************************
```

The library routines are divided into the following groups:
0 - Library utility routines
1 - General utility routines
2 - Solution of nonlinear equations in one variable
3 - Solution of systems of linear equations
4 - Numerical integration.

A summary of each routine is given below; the reader is
referred to the book for further details. References in
square brackets are to the sections of the book where
the routines are discussed and the codes can be found.


```
***********************************
Group 0  Library utility routines
***********************************
```

```
    PROCEDURE residu ( n : range ;  VAR A : matrix ;
                          VAR x, b, r : rvector ) ;
```
External FORTRAN subroutine to compute the residual vector
r = b - Ax for an nxn matrix A and vectors x, b of order n.
The array dimensions must be equal to upbnd (currently 100).
The calculation of inner products and final subtraction
should be performed in double precision.
[7.3]


```
    PROCEDURE errormessage (
                routine : name ;  errornum : integer ;
                errorsummary : message ;  condition : severity ;
                VAR eflag : integer ) ;
```
Handles error messages and warnings in the mathlib library.
A message is written to the output file if eflag=0 on entry.
If condition=fatal, an error message is issued;
if condition=warning, a warning is issued.
The output consists of the name of the routine (routine),
the error number (errornum) and the error message or warning
(errorsummary). Routine and errorsummary must be strings of
length exactly 8 and 30 characters respectively. In all
cases the error number is returned in eflag.
[2.2]


```
    FUNCTION tolmin ( x : real ) :  real ;
```
Computes the minimum tolerance to be used in error tests.
[4.3]


```
    FUNCTION mult ( x, y : real ) :  real ;
```
Computes the product x*y, avoiding underflow failure.
[7.3]


```
    FUNCTION inprod ( VAR A : matrix ;  VAR x : rvector ;
                        i, lower, upper : range ) :  real ;
```

Computes the inner product A[i,j]*x[j], j = lower,...,upper,
avoiding underflow failure in individual products.
Routine called from the library: mult.
[7.3]

FUNCTION longinprod (n   : longrange ;
                     VAR x, y : longvector) :   real ;
Computes the inner product x[j]*y[j], j = 1,...,n,
for long vectors x and y, avoiding underflow failure.
Routine called from the library: mult.
[7.6]

FUNCTION stepmin (a, b : real) :   real ;
Computes the minimum stepwidth to be used for numerical
integration in the interval [a, b].
Routine called from the library: tolmin.
[9.3]

****
Group 1   General utility routines
****

FUNCTION sign (x : real) :   integer ;
Returns the sign of x   (sign = 1 if x is positive,
= -1 if x is negative, = 0 if x is zero).
[1.4]

FUNCTION powerr (x, r : real) :   real ;
Computes x to the power r, where r is real.
Returns 0.0 in cases where x^r would underflow.
Writes an error message and causes runtime failure
if x^r is undefined.
Routine called from the library: mult.
[1.4, app.C]

FUNCTION poweri (x : real ; i : integer) :   real ;
Computes x to the power i, where i is an integer.
Returns 0.0 in cases where x^i would underflow.
Writes an error message and causes runtime failure
if x^i is undefined.
Routine called from the library: mult.
[1.4, app.C]

PROCEDURE graf (VAR fileout : text ;   npts : range ;
                VAR x, y   : rvector) ;
Plots the graph of the points (x[i], y[i], i = 1,...,npts),
where npts <= upbnd. Graph is written to the file fileout.
No auxiliary routines are called.
[1.6, 2.3]

FUNCTION polsum (n : index ;   VAR a : coeff ;   x : real)
                : real ;
Evaluates the polynomial of degree n >= 0,
   a[0] + a[1]*x + ... + a[n]*x^n,
for a specified value of x using Horner's scheme.
[2.2, 2.3]

```
PROCEDURE polval ( n : index ;  VAR a : coeff ;  x : real ;
                   VAR val, grad : real ) ;
Computes the value (val) and the derivative (grad) of the
polynomial
        a[0] + a[1]*x + ... + a[n]*x^n
for a specified value of x and n >= 0 using extended
Horner's scheme.
[3.3]


FUNCTION roundt ( x : real ) :  real ;
Returns x rounded to 4 significant decimal digits.
Routines called from the library:  sign, poweri (mult).
[4.4, app.C]
```

```
************************************************************
Group 2  Solution of nonlinear equations in one variable
************************************************************
```

```
**************************
2.1  Polynomial equations
**************************
```

The following routine computes an approximation to a real root
of a polynomial equation:

```
PROCEDURE polnewt (
              n : index ;  VAR a : coeff ;  tol : real ;
              trace : boolean ;  VAR root : real ;
              VAR b : coeff ;  VAR eflag : integer ) ;
Computes a real root of the polynomial equation
      a[0] + a[1]*x + a[2]*x^2 + ... + a[n]*x^n = 0,
using Newton's method with polval to evaluate the polynomial
and its derivative. The user must supply an initial estimate
in the parameter root. The iteration terminates when the
correction to the computed root is less than
        tol + tolmin(root)
or limiting precision is reached. The remaining
roots can be determined from the deflated polynomial
      b[0] + b[1]*x + b[2]*x^2 + ... + b[n-1]*x^(n-1),
where b[0], ..., b[n-1] are returned in the array b.
A trace of the iteration can be obtained by calling the
routine with trace=true.
The error number for eflag is:
  =1  Newton's method fails to converge after
      20 iterations  (fatal).
See above, "Error messages and warnings", for general
information on eflag.
Routines called from the library:
polval, tolmin, errormessage.
[5.4]
```

```
*********************************
2.2  General nonlinear equations
*********************************
```

This group of routines uses interval iterative methods to
compute a root of the equation f(x)=0 in an interval [a, b],
where f(a)*f(b) <= 0. The function f, which must be continuous
on [a, b], is supplied by the user as a Pascal function:
```
        FUNCTION f ( x : real ) :  real
```
Three routines are included with the same parameter list
and specifications:

```
PROCEDURE <name> ( FUNCTION f(x:real) : real ;   a, b : real ;
                   tol : real ;   VAR root : real ;
                   VAR eflag : integer ) ;
```
The iteration is terminated when the error in the computed
root is less than
```
         tol + tolmin(root)
```
or limiting precision is reached.
The error numbers for eflag are:
```
   =1  The function does not change sign over [a, b],
       i.e.  f(a)*f(b) > 0  (fatal)
   =2  Limiting precision is reached,
       i.e.  abs(f(root)) < smallreal,
       before the convergence test is satisfied. The
       latest estimate is returned in root  (warning).
```
Routines called from the library:
sign, errormessage, tolmin.

The three routines included are:

```
PROCEDURE bisect
Simple bisection method
[6.1]

PROCEDURE bisecant
Bisection/secant method
[6.3]

PROCEDURE bisrat
Bisection/rational interpolation method
(Bus and Dekker algorithm)
[6.5]
```

```
***************************************************
Group 3  Solution of systems of linear equations
***************************************************
```

```
****************************
3.1  General dense matrices
****************************
```

```
PROCEDURE lufac ( n : range ;   VAR A, LU : matrix ;
                  VAR row : ivector ;   VAR eflag : integer ) ;
```
Factorises a nonsingular nxn matrix A (with row permutation)
in the form A' = L*U, where L is a unit lower triangular
matrix and U is an upper triangular matrix. The routine uses
Gaussian elimination with scaling and partial pivoting.
The triangular factors, minus the unit diagonal of L, are
returned in the array LU. The permuted row indices are
returned in the vector row. Calling lufac (n, A, A, row,
eflag) will overwrite A with LU.
The error number for eflag is:
```
   = 1  Matrix is singular to within rounding error  (fatal).
```
Routine called from the library: errormessage.
[7.2]

```
PROCEDURE lusolv ( n : range ;   VAR LU : matrix ;
                   VAR b : rvector ;   VAR row : ivector ;
                   VAR x : rvector ) ;
```
Computes an approximate solution of the nxn linear system
Ax = b, where b is a given vector. The routine must be
preceded by lufac, which prepares the triangular factors of
A in the array LU and records the permuted row indices in
row. The system L(Ux) = b' is solved by forward and backward

substitution, and the solution is returned in x.
There are no error numbers in lusolv.
Routines called from the library: inprod (mult).
[7.3]

PROCEDURE linsolv (n : range ;   VAR A, LU : matrix ;
                    VAR b : rvector ;   VAR row : ivector ;
                    VAR x : rvector ;   VAR eflag : integer) ;
Computes an accurate solution of the nxn linear system
Ax = b, where b is a given vector. The routine must be
preceded by lufac, which prepares the triangular factors of
A in the array LU and records the permuted row indices in
row. The system is solved by substitution and iterative
refinement, and the solution correct to machine accuracy
is returned in x.
The error number for eflag is:
  =1 Iterative refinement fails to converge - matrix is
      too ill-conditioned for system to be solved (fatal).
Routines called from the library:
lusolv (inprod (mult)), tolmin, errormessage.
**NOTE: an external FORTRAN subroutine RESIDU is called to
compute accurate residuals - see comment at head of library.
[7.3]

****
3.2 Large sparse matrices (positive definite)
****

PROCEDURE congrad (n : longrange ;
                    PROCEDURE aprod (n:longrange;
                                     VAR y,z:longvector) ;
                    VAR b, x : longvector ;
                    VAR count, eflag : integer) ;
Solves the nxn linear system Ax = b by conjugate gradient
method. The matrix A must be symmetric and positive definite
and is normally large and sparse. The matrix is not stored
explicitly; instead, a procedure must be supplied in the
calling program corresponding to the formal parameter aprod.
The procedure should compute z = Ay for any given vector y.
After successful call of congrad, x contains an approximate
solution of Ax = b, count records the number of iterations.
The error number for eflag is:
  =1 Matrix is not positive definite (fatal).
Routines called from the library:
longinprod, mult, errormessage.
[7.6]

****
Group 4 Numerical integration
****

****
4.1 Fixed-point methods
****

PROCEDURE simpint (n : range ;   h : real ;   VAR y : coeff ;
                    VAR ivalue : real) ;
Estimates the integral of a function whose numerical values
y[0], ..., y[n], n <= upbnd, are given at equally-spaced
points with stepwidth h. Composite Simpson's rule is used,
with one application of the 3/8 rule if n is odd (if n=1 the
trapezoidal rule is used). Computed value of the integral is

returned in ivalue. There are no error numbers and no
indicator of accuracy in simpint.
[8.2]

PROCEDURE gaussint (FUNCTION f(x:real) : real; a, b : real;
                     p : gaussrules ; n : posint ;
                     VAR ivalue : real) ;
This routine estimates the integral of a function f(x)
from x=a to x=b using a composite p-point Gaussian rule,
where p is an integer from 1 to 4 selected by the user.
The interval [a, b] is partitioned into n subintervals
of equal width, and the p-point Gaussian rule is used to
estimate the integral of f over each subinterval. The
values of f at the limits a and b are not used.
The computed value of the integral is returned in ivalue.
There are no error numbers and no indicator of accuracy
in gaussint.
[8.3]

****
4.2 Adaptive methods
****

The routines in this group may be used to estimate the
integral of an analytically-defined function f(x) from
x=a to x=b to a specified absolute error tolerance tol.
The following parameters are common to all the routines:

    FUNCTION f(x:real) : real ;   a, b : real ;   tol : real ;
    VAR ivalue, errbnd : real ; VAR eflag : integer

The computed value of the integral and an estimated error
bound are returned in ivalue and errbnd respectively.
The process terminates normally when
              errbnd < tol.
For nonstandard terminating conditions and error numbers, see
under the specification of each routine below.

Because no automatic integration process is equally efficient
over a range of different integrals, two integrators based on
different methods are provided. The methods are:
    (1) Romberg integration. This is a whole-interval process
        in which the interval [a, b] is partitioned into
        subintervals of equal width. The stepwidth for the
        required accuracy is determined automatically by the
        routine. This method is most efficient if high accuracy
        is required and the integrand f is a smooth function;
        the method can be very inefficient otherwise.
    (2) Adaptive Simpson's rule (Mckeeman/Lyness algorithm).
        The interval [a, b] is partitioned recursively into
        subintervals, usually of unequal width, until sufficient
        accuracy can be obtained by two applications of
        Simpson's rule on each subinterval.

The two routines which implement these methods are:

PROCEDURE romberg (
            FUNCTION f(x:real) : real ;   a, b, tol : real ;
            trace : boolean ; VAR ivalue, errbnd : real ;
            VAR eflag : integer) ;
This uses Romberg integration with de Boor's cautious
extrapolation method to test for satisfactory convergence.
If a reliable error bound cannot be determined, errbnd is
returned as 1/smallreal and eflag is set to 2. If trace=true
on entry, the Romberg table and convergence ratios are
written to the output file.

The error numbers for eflag are:
=1 The required accuracy has not been attained -
    either limiting precision of the computer has been
    reached or the maximum 15 rows of the Romberg table
    have been used. An estimated error bound is still
    returned but this lies outside the specified
    tolerance (warning)
=2 No reliable error bound has been determined (fatal).
Routines called from the library:
errormessage, tolmin, stepmin.
[9.3]

PROCEDURE adsimp (parameter list
                    same as in romberg) ;
This uses an adaptive integration scheme based on Simpson's
rule. The recursive behaviour of the algorithm can be
exhibited by setting trace=true on entry.
The error numbers for eflag are:
=1 The required accuracy has not been attained -
   limiting precision of the computer reached (warning)
=2 The required accuracy has not been attained -
   maximum depth of recursion reached (warning).
Routines called from the library:
errormessage, tolmin, stepmin.
[9.4]

The above two routines can be accessed independently or via
the master integration routine integral; this also provides a
set of transformations to extend the class of integral which
can be handled by the two integrators.

PROCEDURE integral (
        FUNCTION f(x:real) : real ; a, b, tol : real ;
        routine : selectint ; interval : intertype ;
        transform : boolean ; VAR ivalue, errbnd : real ;
        VAR count, eflag : integer) ;
This routine estimates the integral of f(x) between the
limits a and b, either or both of which may be infinite.
The user selects the integrator by assigning one of the
following values to the parameter routine:
   rombergint - Romberg integration (PROCEDURE romberg)
   adsimpint - adaptive Simpson's rule (PROCEDURE adsimp).
The type of interval is specified by the value of interval:
   AtoB   - the integral is over a finite interval [a, b].
   AtoINF   - the integral is over a semi-infinite interval
                [a, inf). The parameter b is not used.
   MINFtoB   - the integral is over a semi-infinite interval
                (-inf, b]. The parameter a is not used.
   MINFtoINF- the integral is over the infinite interval
                (-inf, inf). The parameters a, b are not used.
If the parameter transform=true, the integral is transformed
to one over [0, 1] by the "HP transformation" x = T(y).
This option should normally be used in any of the following
cases:
   (1) the integrand is defined only in the limit at a or b
       (f(x) need not be specified at such points);
   (2) the integrand has a mild singularity at a or b;
   (3) the derivative of the integrand is singular at a or b;
   (4) one or both of the limits of integration is infinite;
   (5) a consistency check is required between the results
       produced with and without the transformation (the
       actual integration points used are different).

[Note: the HP transformation will remove a singularity in
the integrand at c = a or b provided there exists a real
number p <= 1/2 such that
     lim (f(x)*(abs(x-c))^p) = 0 as x tends to c.
For infinite limits of integration, valid results are

assured provided there exists a real number q >= 3/2
(or q >= 2 if the HP transformation is not used) such that
    lim (f(x)*(abs(x))^q) = 0   as x tends to inf or -inf.
The transformation may also be tried, with extreme caution,
if the conditions are not known to be satisfied. For cases
where the transformation will remove a singularity in f'(x)
at a or b, see section 9.5 of the book.]

The parameter count returns the number of evaluations of the
integrand. The other parameters are the same as in romberg
and adsimp.
The error numbers for eflag are:
  =1   corresponds to eflag=1 in routine selected by routine
  =2   .   .   .   eflag=2 .   .   .   .
Routines called from the library:
either romberg (errormessage, tolmin, stepmin)
or   adsimp (errormessage, tolmin, stepmin).
[9.5]   *)

(* ********************
   The library routines
   ******************** *)

<The code of the library routines should be inserted here>

(* ******************
   Dummy main program
   ****************** *)

BEGIN
END .

Appendix C: The Utility Functions powerr, poweri and roundt

```
FUNCTION powerr ( x, r : real ) :  real ;
(* Computes x to the power r, where r is real.
   Returns 0.0 in cases where x^r would underflow.
   Writes an error message and causes runtime failure
   if x^r is undefined. Avoids underflow failure by using
   the library constant smallog and the function mult *)

   VAR
      lnxr : real ;


   PROCEDURE xrundefined ;
   (* Locally-declared procedure invoked by powerr when x^r is
      undefined, i.e. when (x < 0)   OR  ((x = 0) AND (r <= 0)) *)

      VAR
         zero : integer ;

      BEGIN
      writeln ;  write ('*** error in powerr :  ') ;
      IF x < 0.0 THEN
         writeln ('x < 0')
      ELSE
         writeln ('x = 0, r <= 0') ;
      zero := 0 ;  x := 1/zero  (* force runtime failure *)
      END (* xrundefined *) ;


   BEGIN  (* body of powerr *)
   IF x > 0.0 THEN
      BEGIN
      lnxr := mult(r, ln(x)) ;
      IF lnxr > smallog THEN
         powerr := exp(lnxr)
      ELSE
         powerr := 0.0  (* x^r may underflow *)
      END
   ELSE
      IF x = 0.0 THEN
         IF r > 0.0 THEN
            powerr := 0.0
         ELSE
            xrundefined  (* x = 0, r <= 0 *)
      ELSE
         xrundefined  (* x < 0 *)
   END (* powerr *)
```

Note

powerr handles underflow in x^r as follows: if $r*\ln(x) \leqslant \ln(\text{smallreal})$, that is, $x^r \leqslant \text{smallreal}$, then powerr is set to 0.0. The library constant smallog = ln (smallreal) is used in this test. As a further precaution, $r*\ln(x)$ is evaluated by the library function mult, which avoids possible underflow failure in computing the product.

```
FUNCTION poweri ( x : real ;
                  i : integer ) :  real ;
(* Computes x to the power i, where i is an integer.
   Returns 0.0 in cases where x^i would underflow.
   Writes an error message and causes runtime failure
   if x^i is undefined. Avoids underflow failure by
   using the library function mult *)

   VAR
      power : real ;
      k     : integer ;
      op    : ivector ;

   PROCEDURE xiundefined ;
   (* Locally-declared procedure invoked by poweri when x^i is
      undefined, i.e. when ((x = 0) AND (i <= 0)) *)

      VAR
         zero : integer ;

      BEGIN
      writeln ;
      writeln ('*** error in poweri :  x = 0, i <= 0') ;
      zero := 0 ;  x := 1/zero  (* force runtime failure *)
      END (* xiundefined *) ;

   BEGIN  (* body of poweri *)

   IF x <> 0.0 THEN
      BEGIN
      IF i = 0 THEN
         poweri := 1.0
      ELSE
         BEGIN  (* check for i negative *)
         IF i < 0 THEN
            BEGIN
            x := 1/x ;  i := abs(i)
            END ;
         (* Record in op the required sequence of operations *)
         k := 0 ;
         WHILE i > 1 DO
            BEGIN
            k := k + 1 ;
            IF odd(i) THEN
               op[k] := 1    (* squaring and multiplication *)
            ELSE
               op[k] := 0 ;  (* squaring only *)
            i := i DIV 2
            END (* i *) ;
         (* Compute the power using operations recorded in op *)
         power := x ;
         WHILE k > 0 DO
            BEGIN
            power := mult(power, power) ;
            IF op[k] = 1 THEN
               power := mult(power, x) ;
            IF power = 0.0 THEN
               k := 0
            ELSE
               k := k - 1
            END (* k *) ;
         poweri := power
         END
      END  (* x <> 0.0 *)
```

```
ELSE  (* x = 0.0 *)
   IF i > 0 THEN
      poweri := 0.0
   ELSE
      xiundefined  (* x = 0, i <= 0 *)

END (* poweri *)
```

Note

poweri evaluates x^i, for $x \neq 0$ and $i \neq 0$, by a sequence of squaring and multiplication operations. The required operations are determined from $|i|$ and coded as binary digits in the vector op; 1 denotes squaring followed by multiplication, 0 denotes squaring only. For example, with $i = 13$ the operations vector would be $(1\ 0\ 1)$, and x^{13} would be evaluated as $\{[(x^2)x]^2\}^2 x$. Underflow failure is avoided by computing all squares and products with the library function mult.

```
FUNCTION roundt ( x : real ) :  real ;
(* Returns the value of a real number x rounded to
   4 significant decimal digits. Routines called
   from the library:  sign, poweri (mult) *)

   CONST
      t = 4 ;  (* number of significant digits required *)

   VAR
      scale, power : real ;

   BEGIN
   IF (x = 0.0) OR (t <= 0) THEN
      roundt := 0.0  (* rounded value is zero *)
   ELSE
      BEGIN
      (* Remove and save the sign of x *)
      scale := sign(x) ;  x := abs(x) ;
      (* Scale x to lie between 10^(t-1) and 10^t *)
      power := poweri(10.0, t-1) ;
      IF x < power THEN
         REPEAT  (* scale x up *)
            x := 10*x ;  scale := 10*scale
         UNTIL x >= power
      ELSE
         WHILE x > 10*power DO
            BEGIN  (* scale x down *)
            x := x/10 ;  scale := scale/10
            END ;
      (* Round x to nearest integer using the predeclared
         function round, then scale back and restore sign *)
      roundt := round(x) / scale
      END
   END (* roundt *)
```

Note

To round x to t significant decimal digits, where $t > 0$, x is scaled by a power of 10 to satisfy $10^{t-1} \leqslant |x| \leqslant 10^t$. The result is rounded to the nearest integer by the Pascal function round and finally scaled back by the inverse of the original scale factor.

Appendix D: Notes on Portability

Although we have adhered as far as possible to standard Pascal, it is necessary to make minor changes to the syntax of the library and example programs to make them compatible with some Pascal implementations. This is mainly because the library mechanism is not defined in standard Pascal and is therefore implementation-dependent; also there are minor differences between compilers in respect of the case of reserved words and predeclared entities, and in the use of function and procedure names as arguments to procedures.

The version of the library given in this book is close to the form required by the P4 Pascal compiler to run under the CMS operating system on the Amdahl 470 V/7 (the same compiler also runs on the IBM 370 range under CMS.) The necessary changes are

(i)　Pascal predeclared entities (for example, the standard data types real, integer etc., and the predeclared functions and procedures) must be in upper case.

(ii)　When a Pascal function f(x) is passed as a parameter to a procedure, the argument x and its type must not be specified. For example, the heading of the mathlib routine bisect is given in the text as

```
bisect (FUNCTION f(x:real) : real ;  ...
```

For the Amdahl this must be written as†

```
bisect (FUNCTION f : REAL ; ...
```

The mathlib routines bisect, bisecant, bisrat, gaussint, romberg, adsimp, integral and their respective example programs require this change. Similarly, when a procedure name is passed as a parameter the arguments should not be specified, and hence the arguments to the procedure aprod in the heading of congrad must be deleted.

(iii)　It is necessary to include the compiler switch W− at the top of each example program and at the top of mathlib. The switch is necessary because without it programs using non-standard features of Pascal (for example EXTERN) will not compile.

```
bisect (FUNCTION f(real) : real ;  ...
```

†Some compilers, for example, the DEC system-10 compiler, use the alternative form

275

(iv)   If the mathlib procedure linsolv is required, it will be necessary to load the FORTRAN module RESIDU. To make RESIDU compatible with IBM FORTRAN it is necessary to replace REAL and DOUBLE PRECISION by REAL*8 and REAL*16 respectively.

The Amdahl/IBM version of the library is available from the Numerical Algorithms Group (see preface).

Use of mathlib on Other Machines and Compilers

It is not possible to give full details of how to mount and run the library on different machines; the reader should consult the computer documentation. Below we mention some of the points which should be borne in mind.

The main difference is in the syntax of the Pascal library; this affects both the heading and the dummy program at the end of the library. For example, the DECsystem-10 requires the heading to be of the form

```
PROGRAM mathlib , <list of the mathlib routines
                   separated by commas> ;
```

There is a limit to the number of external routines which can be specified in any one library. It may therefore be necessary to partition the library. This can best be done by splitting the library into its component groups; only the routines in groups 0 and 1 are required by routines in the other groups. In addition, some compilers require that the library terminates after the last routine has been declared.

A number of modern compilers (for example, Berkeley Pascal running under the UNIX operating system, and the UCSD Pascal compiler) treat external declarations rather like forward procedure declarations. An example of this is given below for the Apple II computer.

Finally, the machine-dependent constants will need to be changed if the library is used on a machine that is not compatible with the IBM 370 range.

To use the complete library it is necessary to load a FORTRAN module RESIDU. The only library routine that makes use of this module is linsolv; residu and linsolv may therefore be deleted from the library if it is inconvenient (or impossible) to load the library with a FORTRAN module.

Microprocessors – Apple II

There is no reason why the library cannot be run on a microprocessor provided that a Pascal system which supports libraries is available. However, because of the limitations on storage it is advisable to partition the library into separate groups.

Subsets of the library are available for the Apple II computer (see preface). As shown in the example below, the library mechanism used by the Apple Pascal system (based on UCSD Pascal) is very convenient because the user does not need to declare the library types and routines in the calling program. The library of utility routines (a selection from mathlib groups 0 and 1) on the Apple II has the form

```
(*$S+  swapping option *)
UNIT mathuty ;   (* name of the library *)
   (* Contains the mathlib routines:
      mult, poweri, powerr, sign, polsum, errormessage,
      tolmin  *)

   USES transcend ;  (* Uses the Apple II Pascal
                         transcendental function library *)

   INTERFACE
      (* The following constants, types and routines are
         available to the calling program *)

      CONST
         smallreal = 1.0E-35 ;
         rprec4    = 2.3E-7  ;
         smallog   = -80.6   ;
         upbnd     = 100     ;

      TYPE
         <definition of the library data types>

      (* Specification of each library utility routine *)
      FUNCTION mult (x, y : real ) : real ;
      FUNCTION poweri (x : real ;  i : integer ) :  real ;
      <similarly specify the headings of powerr, sign,
       polsum, errormessage and tolmin>

      (* End of INTERFACE section *)

   IMPLEMENTATION  (* the following is "private": everything
                       declared below except for the routines
                       mentioned above is unavailable to
                       the calling program *)

      FUNCTION mult ;    (* only the name of the routine
                             is required                *)
         <insert the body of FUNCTION mult>
         END (* mult *) ;

      <similarly insert the code of each routine>

      (* End of IMPLEMENTATION section *)

   BEGIN
   END.
```

Dividing the library into an INTERFACE part and an IMPLEMENTATION part is an excellent idea. The library constants, types and routines listed in the INTERFACE part can then be used in the calling program if the instruction

```
USES transcend, mathuty
```

is inserted immediately after the program heading, as shown in the example below. There is no need for the library constants, types and external routines to be declared explicitly.

Example program for the Apple II using mathuty library

```
PROGRAM testpowerr ;

USES transcend, mathuty ;

VAR
    x, r : real ;

BEGIN
WHILE NOT eof DO
    BEGIN
    readln (x, r) ;
    writeln (x, r, '    ', powerr (x, r))
    END
END.
```

Solutions to Exercises

1.1 0.1 is not represented exactly in a computer with binary or related base, hence x is almost certainly never equal to 10.0. The loop will not terminate! Instead, use a FOR loop with an integer variable i running from 1 to 100; compute x as 0.1*i.

1.2 n = 10: 1.0010288, 0.63600564, 0.42309077
n = 100: 1.0000103, 0.63629147, 0.42272872
n = 1000: 1.0000001, 0.63629433, 0.42272509
Error $\approx 10^{-3}, 10^{-5}, 10^{-7}$. The absolute value of error is approximately proportional to square of stepwidth.

1.3 $x = 1, -2; x = -1.5$; no roots.
Try $(1/a) x - a = 0$, where a is a little less than the maximum real number on your computer; the program will fail on overflow.

1.4 0.19217991, 1.0, 0.0, 0.0, invalid data, −0.092296, 1.0, 0.0, 0.0, invalid data.

1.5 $0.5^{1/3} = 0.79$ correct to 2 decimal places; solution of $x + \ln x = 0: x = 0.57$ correct to 2 decimal places. Consider $f(x) = x + \ln x$ and show that $f'(x) > 0$ for all $x > 0$.

1.7 $i, -i, 0.5, i$, undefined.

2.1 $p(-2) = 0$. Polynomial can be factorised as $(x + 2)(2x^2 + x - 2)$, hence $x = -2, -1.280776, 0.780776$.

2.2 Theoretically the computation times are in the proportion 50 : 3 : 2. In practice this will be altered a little by 'housekeeping' and overheads. The numerical results in general will differ slightly because each arithmetic operation can introduce a rounding error.

2.3 If the schemes are designed efficiently there is little to choose between them; both require n multiplications, n divisions and n additions.

2.4   Root 0.8. There are two other real roots, at about -1.1 and 2.3.

2.5   Use $x^r = 10^{r \log_{10} x}$ evaluate $y = r \log_{10} x$, then 10^y, and show that the constraints on x and y are satisfied.

3.1   (i) $3.14 < \pi < 3.142$.
(ii) R_m is a finite set, whereas $[0, 1]$ is an infinite set (for example, $1/10 \notin R_m$ on a binary computer).
(iii)   If $c < 1, f(x)$ has a minimum value of $\ln (1 - c)$ at $x = c$; if $c \geqslant 1, f(x)$ is unbounded.
(iv)   For any $y > 0$ there are *two* points $x = \pm \sqrt{y}$ such that $x^2 = y$. Since the correspondence is not unique, the inverse relation is not a function as defined.

3.2 $0.2^{1/5} = 0.7248, \quad 0.2^{1/10} = 0.8513$ correct to 4 significant figures.

3.3 $p(3) = 8, p'(3) = 24$ (note that $a_2 = 0$). If n = 0 the loop is bypassed and val: a[0], grad: = 0.0. If n = 1, val: = a[1] *x + a[0], grad: = a[1].

3.4 $0.8628, 0.7000, 0.7800$.
Bounds: $0.7800 < \cos 36° < 0.8213$ (true value: 0.8090).

3.5   Show that if $p(x) - q(x)$ is not identically zero, then it has $n + 1$ real zeros at $x_0, x_1, \ldots, x_n$, contradicting the stated property. Polynomial of degree $\leqslant 3$: $\quad p_1(x) = x + 1$.

3.6   Use Taylor's series expansion of e^{1+x} and take $n = 3$, then $|\text{error}| \lesssim 0.00001$ for $x = 0.1.$ $e^{1.1} = 3.0042$ correct to 4 decimal places.

3.7 $\ln x = - \left(x + \dfrac{x^2}{2} + \ldots + \dfrac{x^n}{n} \right) - \dfrac{x^{n+1}}{(n + 1)(1 - \xi)^{n+1}}$

where ξ lies between 0 and x. Set $x = - 0.2$, then $\ln 1.2 = 0.18267 \pm 0.0004$, 0.182267 ± 0.000064.
True value: 0.182322; errors $-0.00035, 0.000055$.

3.8   Use $\sin (x + m\pi) = (-1)^m \sin x$, where m is an integer. Choose m so that x lies between $-\pi/2$ and $\pi/2$; thus $666 = -0.0176426 + 212\pi$. Then $\sin 666 = -0.01764$ correct to 5 decimal places.

3.10   Error expressions: $-h^2 f'''(\xi)/6$, where $\xi \in (x_0 - h, x_0 + h)$; $\quad h^2 f'''(\xi)/3$, where $\xi \in (x_0, x_0 + 2h)$.

4.1   Transform x as described in the solution to exercise 3.8 so that $-\pi/2 < x \leqslant \pi/2$. Use tolmin in the convergence test.
$\sin 666 = -0.01764175$ correct to 8 decimal places.

4.3   Product: 0.2691×10^{-1}, absolute error 0.152×10^{-5}, relative error 0.5649×10^{-4}; 5 correct decimal places, 4 significant figures. Quotient: 0.7653×10^2, absolute error 0.0407, relative error 0.0005315; 1 correct decimal place, 3 significant figures.

4.4   Bound on relative error in both cases: $\frac{3}{2}\beta^{1-t}$ (to 1st order). In exercise 4.3, $\beta = 10, t = 4$: relative error bound $= 1.5 \times 10^{-3}$.

4.6   Cancellation error can occur in numerically smaller root; see section 4.4.

4.7   Error bounds: $|\epsilon_x| \times |\ln \bar{y}| + |\epsilon_y| \times (\bar{x}/\bar{y})$,

$\bar{x}^{\bar{y}} [|\epsilon_x| \times (\bar{y}/\bar{x}) + |\epsilon_y| \times |\ln \bar{x}|]$.

$x \ln y = 7.049$ (1, possibly 2 correct decimal places),

$x^y = 85.12$ (no correct decimal places).

4.8   Use backward recurrence starting with $P_N = 0$.
$P_0 = (1 - e^{-\pi})/\pi = 0.304554$.

4.9   The sequence for $x = 1.0$ decreases down to $J_4(1) = 0.0044$, then increases rapidly so that $J_{10}(1) = 7029$. The starting values contain errors of up to 0.5×10^{-4} in fourth decimal place. The recurrence relation is $J_{k+1} = 2kJ_k - J_{k-1}$, $k = 1, 2, \ldots$. The propagated error in J_{k+1} is approximately $2^k k! \, \epsilon_0$, where $|\epsilon_0| \lesssim 0.5 \times 10^{-4}$. For J_{10}, $|\text{error}| \lesssim 10^4$. The algorithm is numerically unstable.

4.10   Exact derivative at $x_0 = 1$ is $e = 2.7182818 \ldots$. The absolute error first decreases in proportion to h^2, in agreement with the truncation error formula which is $O(h^2)$. On an 8-digit calculator the error attains a minimum value at about $h = 10^{-3}$. It then increases again, and the approximation for $h = 10^{-6}$ is not much better than for $h = 10^{-1}$. The reason is that for small h the numerator is small, the function values are calculated to finite precision, cancellation error occurs, and on division by $2h$ this can give a large absolute error.

5.1 $x_{k+1} = (1 - 1/n) x_k + c/(n \, x_k^{n-1})$.
$e^{1/3} = 1.3956, \pi^{1/5} = 1.2573$ correct to 5 significant figures.

5.2   For Newton's method, truncate at $f'(x_k)$ and set $f(\alpha) = 0$. By including the next term in the series, obtain higher-order formula

$$x_{k+1} = x_k - [f_k' - \sqrt{(f_k'^2 - 2f_k f_k'')}]/f_k''$$

5.3 $f^{(m)}(\alpha) = m! \, \Phi(\alpha) \neq 0$.
$F'(\alpha) = 1/m \neq 0$, where $F(x) = f(x)/f'(x)$.

5.4 (i) computes $1/c$. $g'(1/c) = 1/(c^2 + 1) \neq 0$: 1st order.

(ii)   computes $1/c$. $g'(1/c) = 0, g''(1/c) = -2c \neq 0$: 2nd order.

(iii)   computes $c^{1/2}$ for suitable starting value. $g'(c^{1/2}) = 0, g''(c^{1/2}) = 0,$
$g'''(c^{1/2}) = 3/c \neq 0$: 3rd order.

Use method (ii) to compute $1/\pi = 0.31831, 1/e = 0.36788$.

5.5 $-1.084334, 0.811530, 2.272804.$
$-3.024429, -0.914050, 0.877626, 2.060853.$

5.6 Use a relative error test. $V = 1.190, 0.4529, 0.2106$ to 4 significant figures.

5.8 Polynomial has five real zeros: $0.5, 1.0, 2.0, 4.0, 8.0$.

5.9 Show that $g'(\alpha) = 0, g''(\alpha) = f''(\alpha)/f'(\alpha)$, where $g(x) = x - f(x)/f'(x)$.
If $f(x) = (x - \alpha)^m \Phi(x)$, then $g'(\alpha) = 1 - 1/m \neq 0$ for $m > 1$. With $g(x) = x - r$
$f(x)/f'(x), g'(\alpha) = 1 - r/m$; hence $g'(\alpha) = 0$ if $r = m$.

5.10 Use the iteration formula of exercise 5.9 with r = multiplicity of zero;
r must be a parameter of the procedure. Not suitable for a library routine as
multiplicity must be known in advance!

6.1 $[-1, 5]$: function does not change sign.
$[-1, 0.5]$: locates discontinuity at $x = 0$, but for tol $= 0.0$ overflow will occur
(singularity at origin).

6.2 (i) 0.56714329, (ii) 0.73908513, (iii) 1.76322283.

6.3 $x_{k+1} = x_k - (x_k^n - c)^2 / [(x_k^n + x_k - c)^n - x_k^n]$,
$x_{k+1} = x_k - (x_k^n - c)(x_k - x_{k-1})/(x_k^n - x_{k-1}^n)$.

6.5 $1.337236, 7.588631, 13.949208.$
Conjecture: roots are separated by approximately 2π.
Root closest to 50 is 51.760122.

6.6 The modified bisecant should work almost as efficiently as bisrat (see table
at end of section 6.5).

6.7 Discrepancies will arise using tol $= 0.0$ because of different machine
precisions; these affect both accuracy of computed root and number of function
calls.

7.1 (i) $(AB)^T_{ij} = (AB)_{ji} = \sum_k a_{jk} b_{ki} = \sum_k (B^T)_{ik} (A^T)_{kj} = (B^T A^T)_{ij}$.

(ii) $(AB)(B^{-1}A^{-1}) = AIA^{-1} = I$, hence $B^{-1}A^{-1}$ is inverse of AB by definition.

(iii)   Take transpose of $AA^{-1} = I$, then $(A^{-1})^T A^T = I^T = I$, hence $(A^{-1})^T$ is inverse of A^T by definition. If A is symmetric, substitute $A^T = A$, then $(A^{-1})^T = A^{-1}$.

7.2 $\begin{bmatrix} 1 & c & c^2 \\ 0 & 1-c^2 & c(1-c^2) \\ 0 & 0 & 1-c^2 \end{bmatrix}$

The given matrix is symmetric and singular if $c = \pm 1$. Then at least two columns are equal, determinant zero.

7.3   Consider the kth elimination stage, count the number of arithmetic operations, and sum from 1 to n. Pivoting will be necessary in general, but interchanges possible only with rows below the pivot. If the matrix is singular, all elements from the pivot down are zero and elimination cannot proceed.

7.4   The non-zero vector $x = (\lambda_1, \ldots, \lambda_n)^T$ is a solution of $Ax = 0$. Any multiple αx also satisfies $A(\alpha x) = 0$.
The solutions are $\alpha(4, 1, -3)^T$ for any scalar α.

7.5   Determinant of 7×7 matrix: $0.38161428 \times 10^{15}$.

7.6   The procedure should check that the argument of the square root is positive, so that l_{jj} is real and positive; otherwise A is not positive definite.

7.7   The approximate solution will be less accurate if a computer of lower precision is used; similarly the refined solution will contain fewer correct digits. To estimate the precision, use $u \sim [n. \text{ cond } (A)]^{-1} (\| \bar{x} - x\| / \| x \|)$.

7.8   The diagonal elements of L are *not* unit; only the lower triangle need be supplied (since $U = L^T$); the permutation vector row is not used. Iterative refinement can be carried out as usual.

7.9   The elements of H_6 are not all machine numbers; there is a data error in representing these. The computed inverse X is the accurate inverse of a perturbed matrix; because of extreme ill-conditioning, X may be quite different from true H_6^{-1}.

7.10   For triangle inequality

$$\| x + y \|_\infty = \max_i | x_i + y_i| \leqslant \max_i (| x_i| + | y_i|)$$

$$\leqslant \max_i | x_i| + \max_i | y_i| = \| x \|_\infty + \| y \|_\infty$$

$$\| A + B \|_\infty = \max_i \sum_j | a_{ij} + b_{ij}| \leqslant \max_i (\sum_j | a_{ij}| + \sum_j | b_{ij}|)$$

$$\leqslant \max_i \sum_j |a_{ij}| + \max_i \sum_j |b_{ij}| = \| \mathbf{A} \|_\infty + \| \mathbf{B} \|_\infty$$

7.11 (i) Take norms in $\mathbf{x} - \overline{\mathbf{x}} = \mathbf{A}^{-1}\mathbf{r}$ and $\mathbf{A}\mathbf{x} = \mathbf{b}$.
(ii) $\| \mathbf{I} \| = \max_{\mathbf{x} \neq \mathbf{0}} (\| \mathbf{I}\mathbf{x} \| / \| \mathbf{x} \|) = 1$, then take norms in $\mathbf{A}\,\mathbf{A}^{-1} = \mathbf{I}$.

7.12 Write 4×2 system $\mathbf{C}\mathbf{x} = \mathbf{b}$ and derive normal equations $\mathbf{C}^T\mathbf{C}\mathbf{x} = \mathbf{C}^T\mathbf{b}$.
Solution: $\mathbf{x} = (0.993, 1.003)^T$; $\| \mathbf{b} - \mathbf{C}\mathbf{x} \|_2 = 0.111$.

7.13 Multiplications: $3n - 3$; divisions: $2n - 1$; additions/subtractions: $3n - 3$.

7.14 $(\mathbf{B}^T\mathbf{B})^T = \mathbf{B}^T\mathbf{B}$. For any $\mathbf{x} \neq \mathbf{0}$, $\mathbf{B}\mathbf{x} \neq \mathbf{0}$ since $\mathbf{B}$ is non-singular; hence
$\mathbf{x}^T(\mathbf{B}^T\mathbf{B})\mathbf{x} = (\mathbf{B}\mathbf{x})^T\,\mathbf{B}\mathbf{x} > 0$.
  For any $\mathbf{x} \neq \mathbf{0}$ and $\mathbf{A}\mathbf{y} = \mathbf{x}$, $\mathbf{y} = \mathbf{A}^{-1}\mathbf{x} \neq \mathbf{0}$. Consider $\mathbf{x}^T \mathbf{A}^{-1} \mathbf{x}$ and substitute
$\mathbf{x} = \mathbf{A}\mathbf{y}$, thus $\mathbf{y}^T\mathbf{A}^T\mathbf{A}^{-1} \mathbf{A}\mathbf{y} = \mathbf{y}^T\mathbf{A}\mathbf{y} > 0$ since $\mathbf{y} \neq \mathbf{0}$ and $\mathbf{A}$ is positive definite.

7.15 $0.695312, 0.697783, 0.698446$. Discretisation error decreases as mesh size is reduced.

7.16 Same results as in exercise 7.15 but should work rather faster.

8.1 A suitable stepwidth would be $1/1200$.
$\pi = 3.141593$ correct to 6 decimal places.

8.2 Integral: 0.3793; estimated error bound: 0.0003.
Simpson's rule gives 0.3796, with an estimated error bound $\sim 10^{-6}$. This is
much less than the possible data error ($\sim 0.4 \times \frac{1}{2} \times 10^{-4}$). Sufficient accuracy
could be obtained in Simpson's rule using only three integration points:
$x = 0.1, 0.3, 0.5$.

8.3 $d_1 = 2, d_2 = -1, d_3 = 2/3, d_4 = -1/2, d_5 = 2/5, d_6 = -1/3$.

8.5 Integral: 9.53. $|$Truncation error$| < 0.071$, $|$data error$| < 0.045$. Only
first figure in integral can be claimed correct.

8.6 (a) 0.107778, (b) 2.57742, (c) 4.46027.

8.7 $\dfrac{(p!)^4 \, h^{2p}}{(2p + 1)\, [(2p)!]^3}\, (b - a)\, f^{(2p)}(\xi), \quad \xi \in (a, b).$

References

Abramowitz, M., and Stegun, I. A. (1965), *Handbook of Mathematical Functions* (Dover Publications, New York)

Addyman, A. M. (1980), A draft proposal for Pascal, *ACM SIGPLAN Notices*, **15**, no. 4, 1-67

Apostol, T. M. (1957), *Mathematical Analysis* (Addison-Wesley, Reading, Mass.)

Burkill, J. C. (1970), *A First Course in Mathematical Analysis* (Cambridge University Press, Cambridge)

Bus, J. C. P., and Dekker, T. J. (1975), Two efficient algorithms with guaranteed convergence for finding a zero of a function, *ACM Trans. Math. Software*, **1**, 330-45

Cohen, A. M., ed. (1973), *Numerical Analysis* (McGraw-Hill, New York and London)

Conte, S. D., and de Boor, C. (1980), *Elementary Numerical Analysis: An Algorithmic Approach*, 3rd ed. (McGraw-Hill, New York and London)

Courant, R., and John, F. (1965), *Introduction to Calculus and Analysis*, vol. 1 (Interscience Publishers, New York)

Dahlquist, G., and Björck, A. (1974), *Numerical Methods* (Prentice-Hall, Englewood Cliffs, N.J.)

Dongarra, J. J., Moler, C. B., Bunch, J. R., and Stewart, G. W. (1979), *LINPACK Users' Guide* (Soc. Ind. Appl. Math., Philadelphia)

Findlay, W., and Watt, D. A. (1978), *Pascal, An Introduction to Methodical Programming* (Pitman Publishing, London)

Forsythe, G. E., and Moler, C. B. (1967), *Computer Solution of Linear Algebraic Systems* (Prentice-Hall, Englewood Cliffs, N.J.)

Forsythe, G. E., Malcolm, M. A., and Moler, C. B. (1979), *Computer Methods for Mathematical Computations* (Prentice-Hall, Englewood Cliffs, N.J.)

Fox, L. (1964), *An Introduction to Numerical Linear Algebra* (Oxford University Press, Oxford)

George, A., and Liu, J. W. (1981), *Computer Solution of Large Sparse Positive Definite Systems* (Prentice-Hall, Englewood Cliffs, N.J.)

Hennell, M. A., and Delves, L. M., ed. (1980), *Production and Assessment of Numerical Software* (Academic Press, New York and London)

Henrici, P. (1964), *Elements of Numerical Analysis* (John Wiley and Sons, New York and London)

Jacobs, D., ed. (1978), *Numerical Software – Needs and Availability* (Academic Press, New York and London)

Jensen, K., and Wirth, N. (1978), *Pascal User Manual and Report*, 2nd. ed. (Springer-Verlag, Berlin)

Kahan, W. M. (1980), Handheld calculator evaluates integrals, *Hewlett-Packard Journal*, August 1980

Madsen, K. (1973), A root-finding algorithm based on Newton's method, *BIT*, **13**, 71-5

Malcolm, M. A., and Simpson, R. B. (1975), Local versus global strategies for adaptive quadrature, *ACM Trans. Math. Software*, **1**, 129-46

Phillips, G. M., and Taylor, P. J. (1973), *Theory and Applications of Numerical Analysis* (Academic Press, New York and London)

Rice, J. R., ed. (1971), *Mathematical Software* (Academic Press, New York and London)

Rice, J. R., ed. (1977), *Mathematical Software III* (Academic Press, New York and London)

Smith, G. D. (1978), *Numerical Solution of Partial Differential Equations: Finite Difference Methods*, 2nd. ed. (Oxford University Press, Oxford)

Vandergraft, J. S. (1978), *Introduction to Numerical Computation* (Academic Press, New York and London)

Varga, R. S. (1962), *Matrix Iterative Analysis* (Prentice-Hall, Englewood Cliffs, N. J.)

Wilkinson, J. H. (1963), *Rounding Errors in Algebraic Processes* (H.M. Stationery Office, London)

Wilkinson, J. H., and Reinsch, C., ed. (1971), *Handbook for Automatic Computation*, vol. 2: *Linear Algebra* (Springer-Verlag, Berlin)

Wilson, I. R., and Addyman, A. M. (1978), *A Practical Introduction to Pascal* (Macmillan Press, London)

Index

abs, function 261
accumulator 81
adaptability 32, 104ff, 229
addition (computer) 82
adsimp, procedure 242, 245, 270
Amdahl 470 computer 77, 79, 163, 262, 275
Apple II computer 276
aprod, procedure 184
arctan, function 261
arithmetic operations 81
array 18ff
  conformant 40ff
  packed 18
ASCII character set 17, 46

back substitution 143ff, 157
battery testing 37, 133
binary fraction 71
bisecant, procedure 130ff, 137, 267
bisect, procedure 121ff, 133, 137, 267
bisection method 50ff, 120
bisection/Newton method 124
bisection/secant method 129
bisrat, procedure 133ff, 267
boolean, data type 4
bound 48
boundary value problem 173

CASE statement 35, 133
cautious extrapolation 223ff, 227
central difference 173
char, data type 18
Cholesky decomposition 155
chopping 78
chr, function 46

coeff, array type 33, 36, 263
compiler switch 275
complex, data type 20
complexample, program 20
condition number 157, 163, 167ff, 187
congrad, procedure 183, 268
conjugate gradient method 180ff
  implementation 182
CONST definition 3, 7
convergence
  first-order (linear) 115
  order 115
  rate 114
  ratio 223, 226, 236
  second-order (quadratic) 115, 117
convexity 103
cos, function 261
Crout's method 155
cutoff 104

decimal places 73
DECsystem-10 computer 276
derivative 52ff
determinant 154, 186
DIV operator 5
Doolittle's method 155

eof, function 8
equilibration: see scaling (linear equations)
error
  absolute 72, 86, 217
  bound 60, 117, 194
  cancellation 84ff
  control 73
  data 70, 165

Numerical Algorithms Group (NAG)
37
numerical instability 88ff

O notation 66
odd, function 261
ord, function 46
order of approximation 66
ordinal type 4
output file 8
output format 9
overflow 4, 76, 122

parameter
  actual 11
  formal 11
  function 122, 275
  list 34
  procedure 183, 275
  value 12ff
  variable 12ff, 20
Pascal 3ff
pivot 143, 150
pivoting 151, 172
polgraph, program 38ff
polnewt, procedure 107ff, 266
polsum, function 36, 38, 265
polval, procedure 54, 107, 266
polynomial
  curve-fitting 170
  definition 28
  deflated 106, 110ff
  evaluation of 30
  interpolating 59, 169
  Legendre 204
  nested form 29
  orthogonal 204
  zeros 110, 112
portability 32, 275ff
posint, data type 205, 263
poweri, function 14, 265, 273
powerr, function 14, 265, 272
precision 76, 229
  double 159
  higher 155, 159
  limiting 105ff, 243
  relative 78ff, 154, 168

procedure
  concept 11
  declaration 11
  use 13
program structure 3

quad, procedure 12
QUADPACK library 247
quadrature: see integration

range, data type 33, 263
read 8
readln 9
real, data type 5
RECORD 20
relative precision: see precision
relaxation 177
REPEAT statement 5
residu, procedure 159, 264
RESIDU, subroutine 160, 262, 276
residual: see vector
Richardson's extrapolation 215, 218
robustness 31, 102ff, 124, 229
romberg, procedure 225, 231ff, 269
root (of equation) 50, 95ff
  complex 97
  double 97
  multiple 97, 113
  simple 96
round, function 7, 261
rounding: see error
roundt, function 82, 266, 274
rprec, program 79
rprec4, constant 33, 78, 79, 263
rvector, array type 33, 263

scalar type 13
scaling (linear equations) 151
secant method 126ff
selectint, data type 255, 263
sequence 60ff
series 61
  alternating 62
  convergent 62
  Fourier 210
  partial sum 61, 83